AF813070

I Can Do Math Practice Problems KINDERGARTEN

Volume 2: Units 5 - 8

This book belongs to:

____

I Can Do Math Practice Problems KINDERGARTEN

Volume 2: Units 5 - 8

Elizabeth Murphy

Published by CKingEducation

CKingEducation

To contact CKingEducation or the authors about speaking, workshops, or ordering books in bulk, visit www.ckingeducation.com.

ISBN: 978-1-968264-06-2

Lead writer: Elizabeth Murphy
Editor: John Sasko
Series editor and book design: Christine King

Printed in the United States of America

Table of Contents

Track what you have done by checking off items.

I did it!	Unit/Practice/I Can Statement	Page
	Unit 5 Practice 1: I can break apart numbers.	16
	Unit 5 Practice 2: I can break apart pattern block designs.	18
	Unit 5 Practice 3: I can match expressions.	20
	Unit 5 Practice 4: I can break apart numbers.	22
	Unit 5 Practice 5: I can count.	24
	Unit 5 Practice 6: I can write an expression.	26
	Unit 5 Practice 7: I can break apart numbers.	28
	Unit 5 Practice 8: I can match expressions.	30
	Unit 5 Practice 9: I can solve a story problem.	32
	Unit 5 Practice 10: I can match a 10-frame to a number.	34
	Unit 5 Practice 11: I can match a 10-frame to an equation.	36
	Unit 5 Practice 12: I can create a 10-frame to match an equation.	38
	Unit 5 Practice 13: I can make 10.	40
	Unit 5 Practice 14: I can make 10.	42
	Unit 5 Practice 15: I can solve a story problem.	44

Table of Contents

Track what you have done by checking off items.

I did it!	Unit/Practice/I Can Statement	Page
	Unit 5 Fluency Practice	46
	Unit 6 Practice 1: I can count objects in a collection up to 20.	52
	Unit 6 Practice 2: I can count objects in a collection up to 20.	54
	Unit 6 Practice 3: I can count objects in a collection up to 20.	56
	Unit 6 Practice 4: I can count objects in a collection up to 20.	58
	Unit 6 Practice 5: I can count objects in a collection up to 20.	60
	Unit 6 Practice 6: I can count objects in a collection up to 20.	62
	Unit 6 Practice 7: I can make numbers with a full 10-frame and more.	64
	Unit 6 Practice 8: I can make numbers with a full 10-frame and more.	66
	Unit 6 Practice 9: I can match 10-frames to equations.	68
	Unit 6 Practice 10: I can complete an equation.	70
	Unit 6 Practice 11: I can color counters to show an expression.	72
	Unit 6 Practice 12: I can count up to 20 objects.	74
	Unit 6 Practice 13: I can understand 10 ones and some more ones.	76
	Unit 6 Fluency Practice	78

Table of Contents

Track what you have done by checking off items.

I did it!	Unit/Practice/I Can Statement	Page
	Unit 7 Practice 1: I can identify missing shapes.	84
	Unit 7 Practice 2: I can identify more or less.	86
	Unit 7 Practice 3: I can identify shapes.	88
	Unit 7 Practice 4: I can match shapes to expressions.	90
	Unit 7 Practice 5: I can match story problems to equations.	92
	Unit 7 Practice 6: I can make 10.	94
	Unit 7 Practice 7: I can identify shapes.	96
	Unit 7 Practice 8: I can tell heavier or lighter.	98
	Unit 7 Practice 9: I can tell which holds more or less.	100
	Unit 7 Practice 10: I can identify solid shapes.	102
	Unit 7 Practice 11: I can describe shapes.	104
	Unit 7 Practice 12: I can identify solid shapes.	106
	Unit 7 Practice 13: I can identify solid shapes.	108
	Unit 7 Practice 14: I can talk about shapes.	110
	Unit 7 Practice 15: I can talk count solid shapes.	112

Table of Contents

Track what you have done by checking off items.

I did it!	Unit/Practice/I Can Statement	Page
	Unit 7 Practice 16: I can match shapes to pictures.	114
	Unit 7 Fluency Practice	116
	Unit 8 Practice 1: I can tell more or less.	122
	Unit 8 Practice 2: I can count objects.	124
	Unit 8 Practice 3: I can add and subtract.	126
	Unit 8 Practice 4: I can find 1 more or 1 less.	128
	Unit 8 Practice 5: I can write the number that comes next.	130
	Unit 8 Practice 6: I can draw circles to match a number card.	132
	Unit 8 Practice 7: I can write numerals and numeral names.	134
	Unit 8 Practice 8: I can color counters to show an expression.	136
	Unit 8 Practice 9: I can answer questions.	138
	Unit 8 Practice 10: I can find the value of an expression.	140
	Unit 8 Practice 11: I can write an equation to match a 10-frame.	142
	Unit 8 Practice 12: I can tell how many using equations.	144
	Unit 8 Practice 13: I can write an expression.	146

Table of Contents

Track what you have done by checking off items.

I did it!	Unit/Practice/I Can Statement	Page
	Unit 8 Practice 14: I can break apart numbers.	148
	Unit 8 Practice 15: I can tell which is more.	150
	Unit 8 Practice 16: I can make 5.	152
	Unit 8 Practice 17: I can make 10.	154
	Unit 8 Practice 18: I can solve a story problem.	156
	Unit 8 Practice 19: I can make 10.	158
	Unit 8 Practice 20: I can tell more or fewer.	160
	Unit 8 Practice 21: I can break apart numbers.	162
	Unit 8 Fluency Practice	164

Welcome to Kinder!

Kindergarten is all about exploring math by counting and seeing the math around us. You will answer questions that start with, "How many..." You will learn the names of shapes. You will explain your thinking. You are becoming a mathematical thinker!

Unit 5

Composing and Decomposing Numbers to 10

In this unit we will represent numbers in different ways. We will also make and break apart numbers within 10.

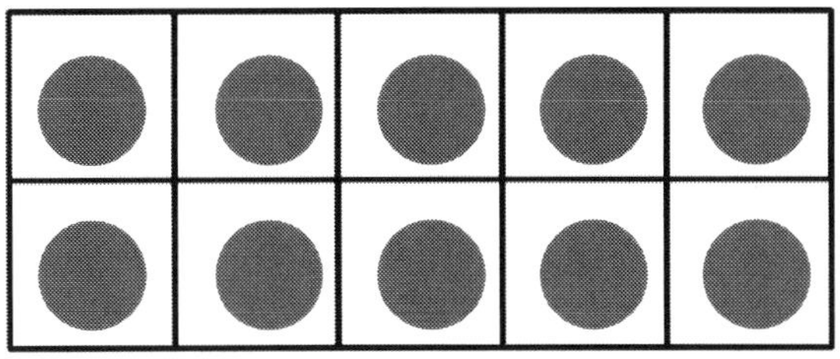

10-Frame

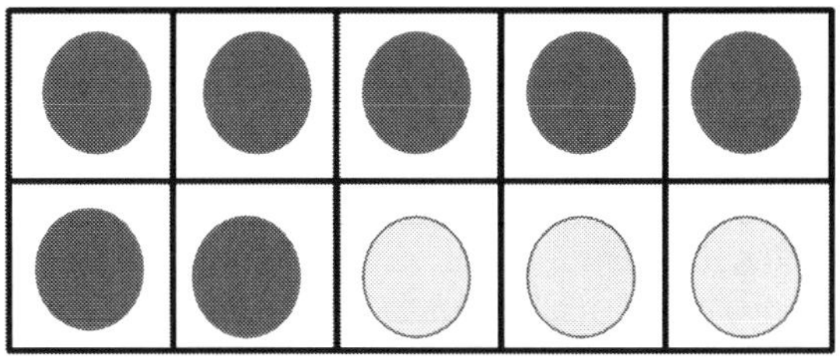

Equation: 10 = 7 + 3

Expression: 3 + 1

4 and 3

Decompose 6

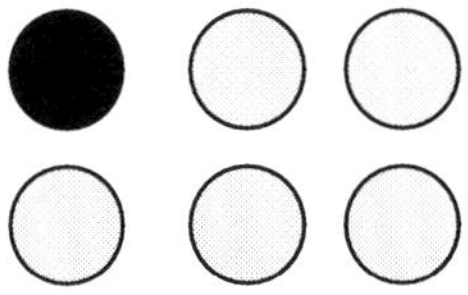

1 and 5

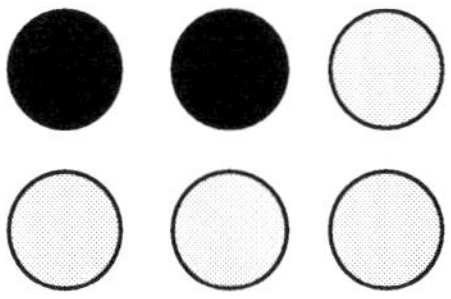

2 and 4

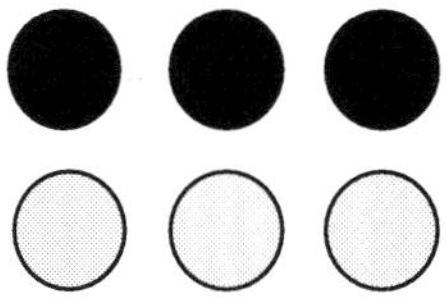

3 and 3

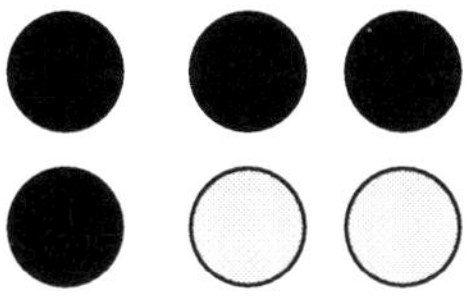

4 and 2

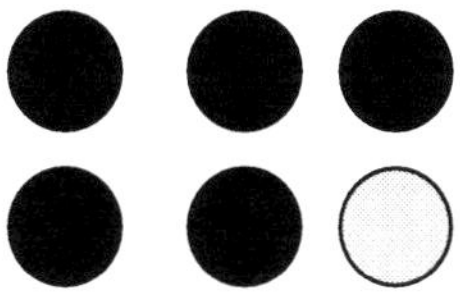

5 and 1

Unit Vocabulary

Composing and Decomposing Numbers to 10

Use this space to visualize the math vocabulary for this unit.

Word or Phrase	Visual Reminder

Unit Tools

Composing and Decomposing Numbers to 10

Use this space to visualize the tools for this unit.

Tools	This is a...	It is used to...

Name: ______________________________ Date: ______________

Unit 5 Practice 1: I can break apart numbers.

1) **Sample**: Break **4** □□□□ squares apart in different ways.

One Way to Make 4	Another Way to Make 4
	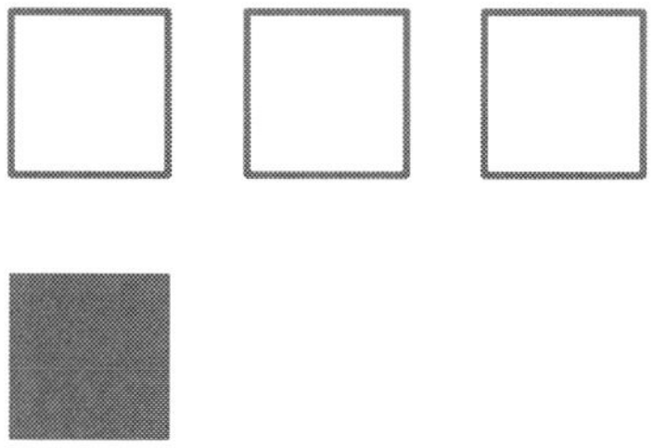
2 + 2	3 + 1

K.OA.3

2) Break **5** □□□□□ squares apart in different ways.

One Way to Make 5	Another Way to Make 5
____ + ____	____ + ____

K.OA.3

Do you understand? **?** ✓

Name: ______________________ Date: ____________

Unit 5 Practice 1: I can break apart numbers.

3) Break **8** ▢▢▢▢▢▢▢▢ squares apart in different ways.

One Way to Make 8	Another Way to Make 8
____ + ____	____ + ____

K.OA.3

4) Break **6** ▢▢▢▢▢▢ squares apart in different ways.

One Way to Make 6	Another Way to Make 6
____ + ____	____ + ____

K.OA.3

Do you understand? **?** ✔

Name: ______________________ Date: __________

Unit 5 Practice 2: I can break apart pattern block designs.

1) **Sample:** Juan used **3** pattern blocks.
Write the expression.

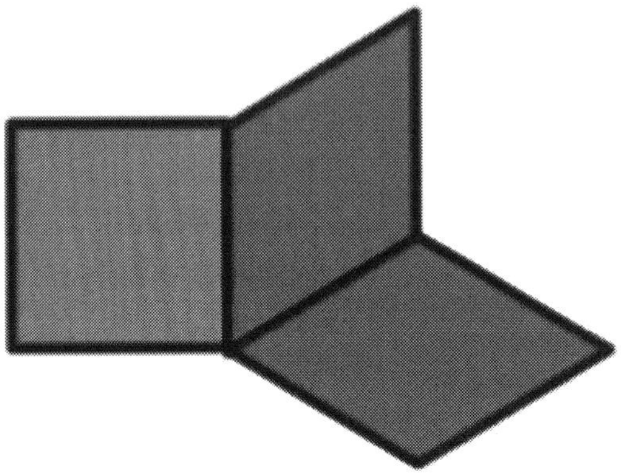

Expression: 1 + 2

K.OA.1

2) Lisa used **4** pattern blocks.
Write the expression.

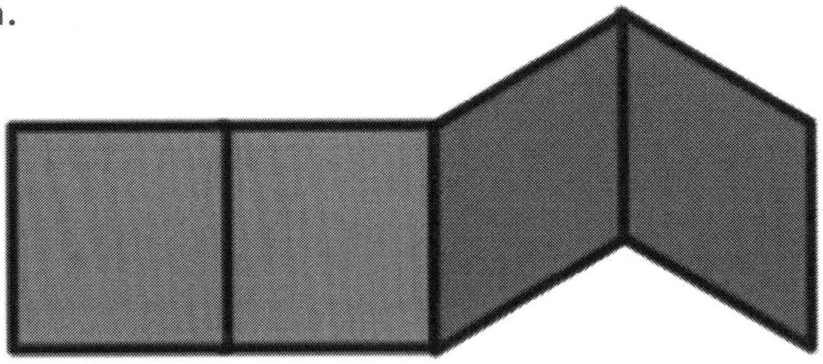

Expression: ______ + ______

K.OA.1

Do you understand? ? ✓

Name: ______________________________ Date: ______________

Unit 5 Practice 2: I can break apart pattern block designs.

3) Diego used **5** pattern blocks.
Write the expression.

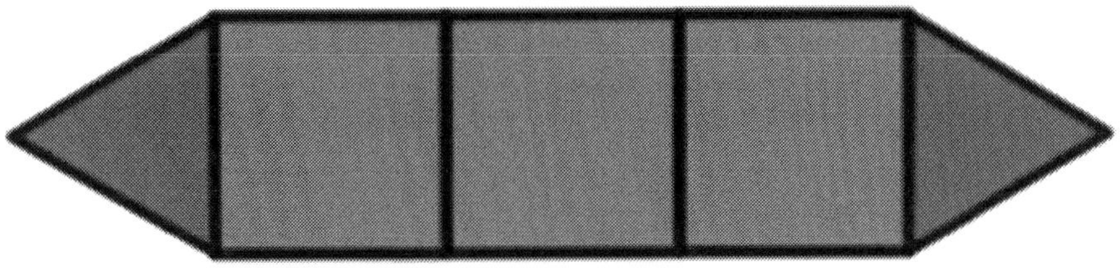

Expression: _____ + _____

K.OA.1

4) Anna used **7** pattern blocks.
Write the expression.

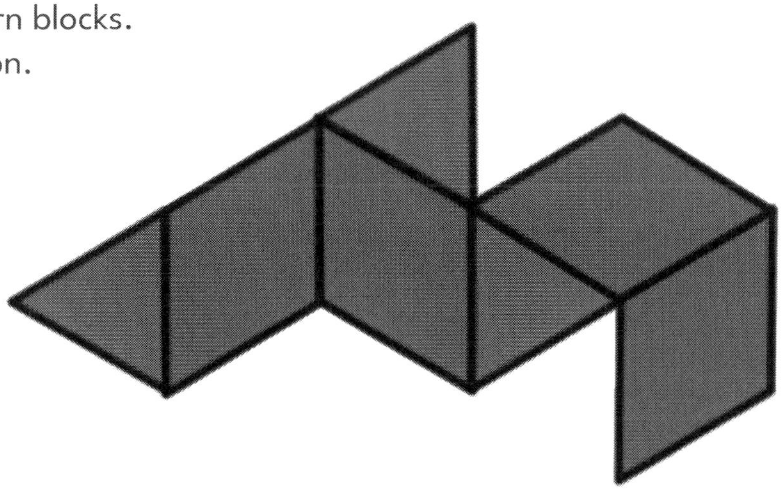

Expression: _____ + _____

K.OA.1

Do you understand? **?** ✓

Name: ____________________ Date: __________

Unit 5 Practice 3: I can match expressions.

1) **Sample:** Color the counters to match the expression.

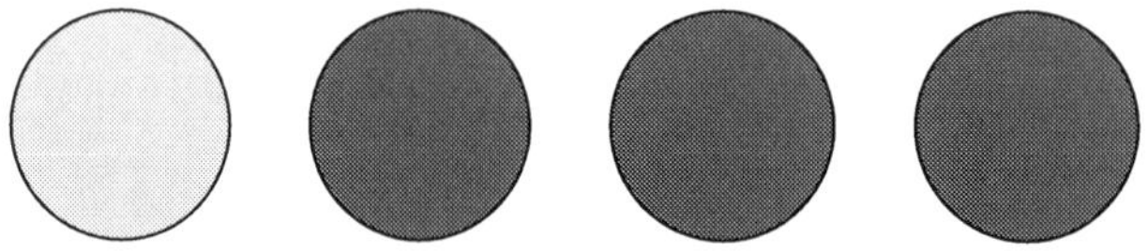

1 + 2

K.OA.3

2) Color the counters to match the expression.

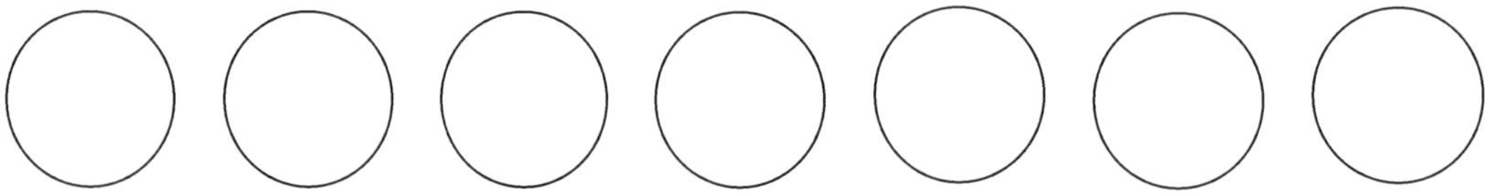

3 + 4

K.OA.3

Do you understand? **?** ✓

Name: ______________________ Date: ______________

Unit 5 Practice 3: I can match expressions.

3) Color the counters to match the expression.

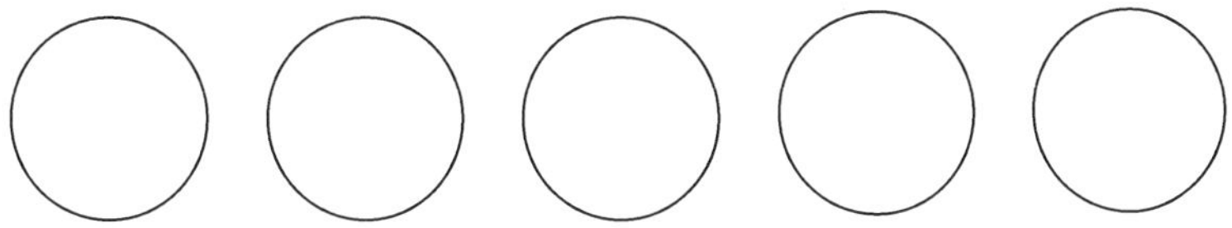

2 + 3

K.OA.3

4) Color the counters to match the expression.

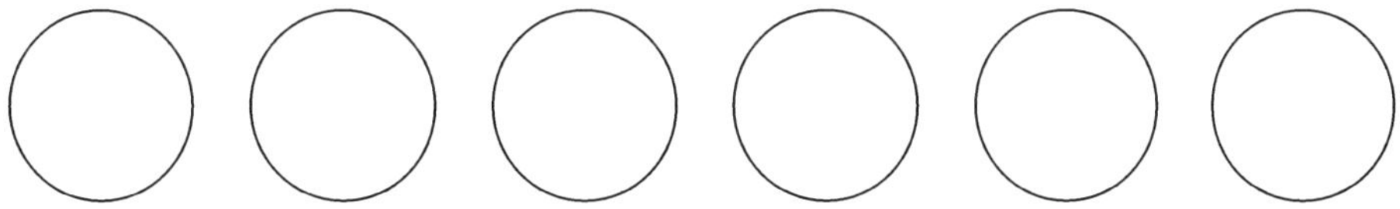

4 + 2

K.OA.3

Do you understand? **?** ✓

Affirmation: Acting out math problems helps me understand.

Name: ______________________________ Date: ______________

Unit 5 Practice 4: I can break apart numbers.

1) **Sample:** Find two ways to break apart **5**. Draw circles to show your work.

2 + 3	1 + 4

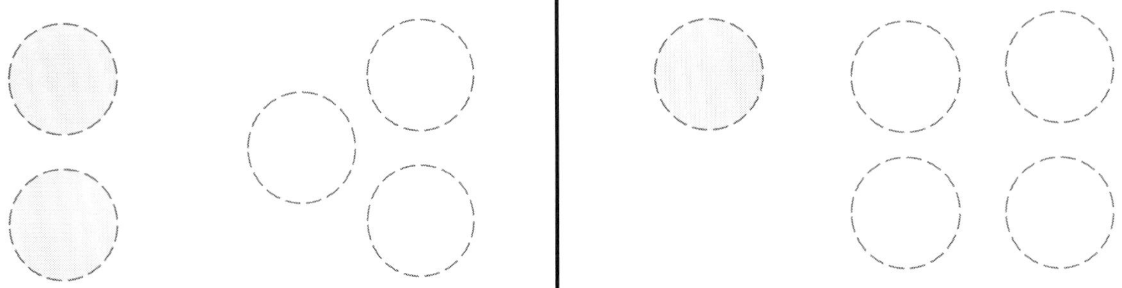

K.OA.3

2) Find two ways to break apart **4**. Draw circles to show your work.

____ + ____	____ + ____

K.OA.3

Do you understand? **?** ✓

Affirmation: Acting out math problems helps me understand.

Name: ______________________ Date: ____________

Unit 5 Practice 4: I can break apart numbers.

3) Find two ways to break apart **8**. Draw circles to show your work.

____ + ____	____ + ____

K.OA.3

4) Find two ways to break apart **7**. Draw circles to show your work.

____ + ____	____ + ____

K.OA.3

Do you understand? **?** ✔

Name: ______________________________ Date: ______________

Unit 5 Practice 5: I can count.

1) **Sample:** How many?

2 + 4

There are 6 pieces of fruit.

K.OA.1

2) How many?

____ + ____

There are ________ pieces of fruit.

K.OA.1

Do you understand? **?** ✓

Affirmation: Acting out math problems helps me understand.

Name: ______________________________ Date: ______________

Unit 5 Practice 5: I can count.

3) How many?

______ + ______

There are ______ pieces of fruit.

K.OA.1

4) How many?

______ + ______

There are ______ pieces of fruit.

K.OA.1

Do you understand? **?** ✓

Name: ______________________________ Date: ______________

Unit 5 Practice 6: I can write an expression.

1) **Sample:** Make believe that the circles are **6** apples. Some of the apples are red and some are yellow. Color the counters. Write the expression.

Expression: 1 + 5

K.OA.1

2) Make believe that the circles are **9** apples. Some of the apples are red and some are yellow. Color the counters. Write the expression.

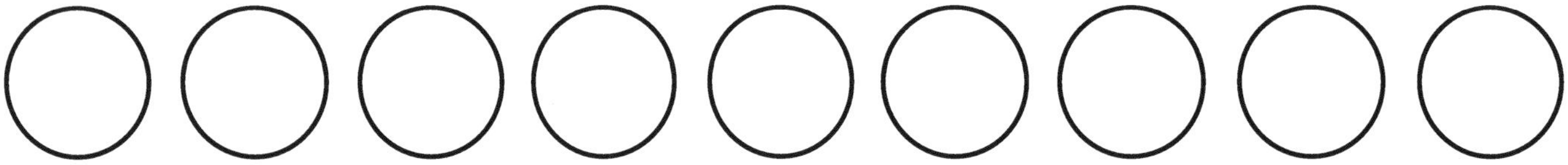

Expression: _____ + _____

K.OA.1

Do you understand? ? ✓

Name: ______________________________ Date: ______________

Unit 5 Practice 6: I can write an expression.

3) Make believe that the circles are **8** apples. Some of the apples are red and some are yellow. Color the counters. Write the expression.

○ ○ ○ ○ ○ ○ ○ ○

Expression: ______ + ______

K.OA.1

4 Make believe that the circles are **7** apples. Some of the apples are red and some are yellow. Color the counters. Write the expression.

○ ○ ○ ○ ○ ○ ○

Expression: ______ + ______

K.OA.1

Do you understand? **?** ✓

Name: ______________________ Date: ____________

Unit 5 Practice 7: I can break apart numbers.

1) **Sample:** Circle the groups that make **5**.

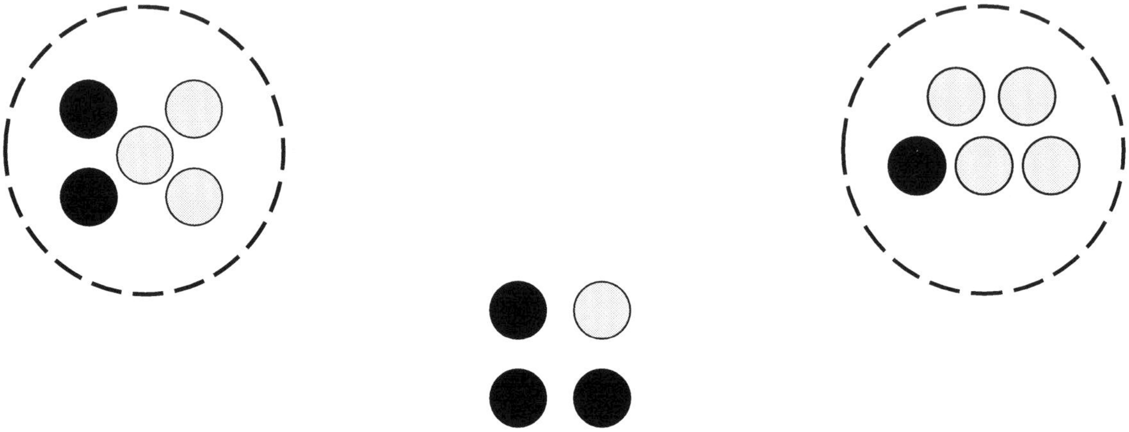

K.OA.3

2) Circle the groups that make **3**.

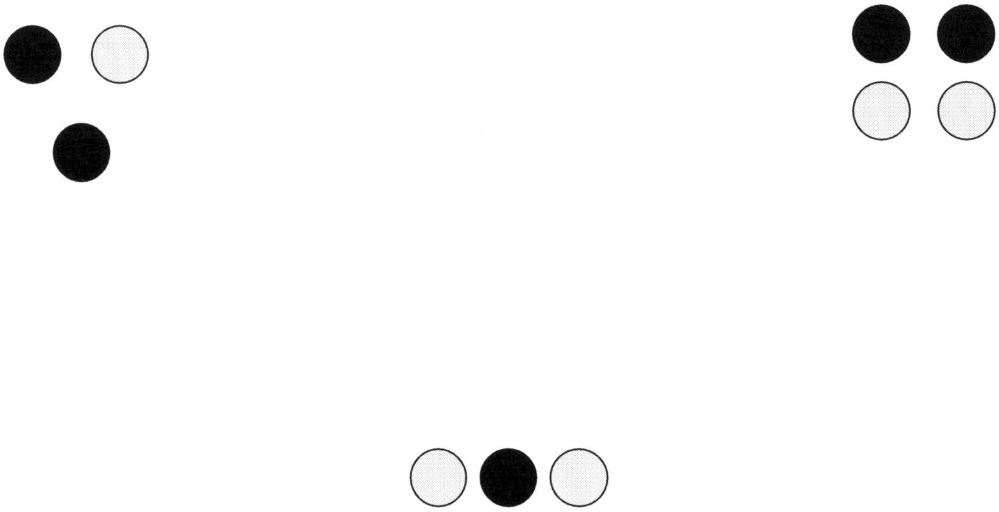

K.OA.3

Do you understand? **?** ✓

Affirmation: Acting out math problems helps me understand.

Name: ______________________________ Date: ______________

Unit 5 Practice 7: I can break apart numbers.

3) **Sample:** Circle the groups that make **6**.

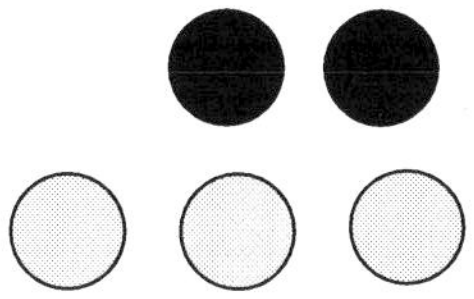

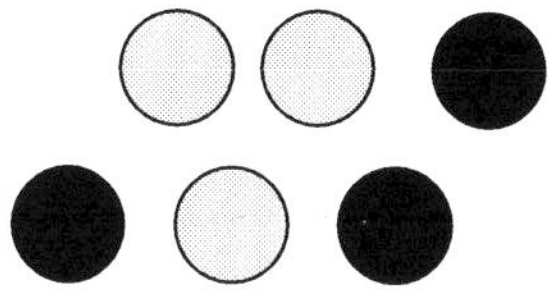

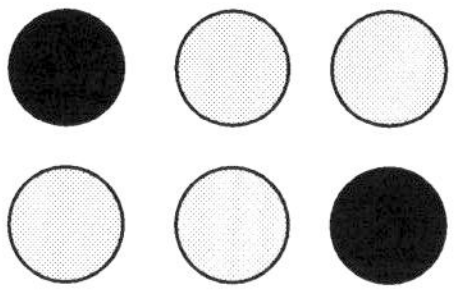

K.OA.3

4) Circle the groups that make **4**.

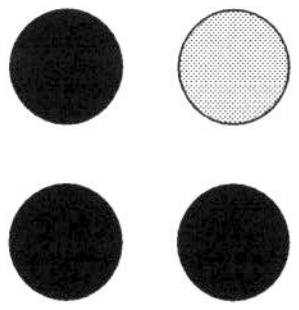

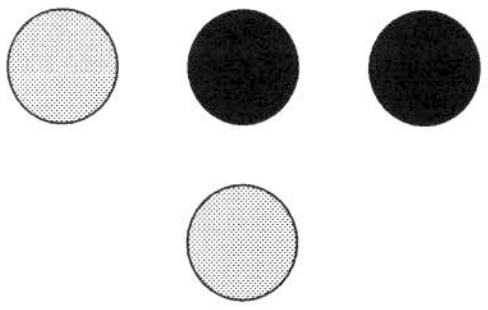

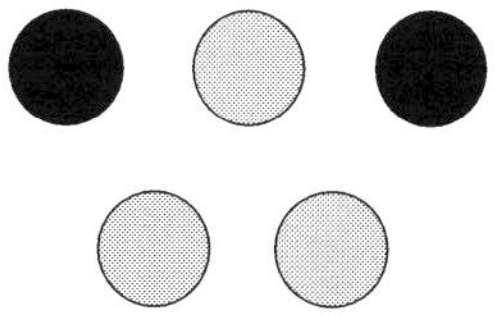

K.OA.3

Do you understand? **?** ✓

Affirmation: Acting out math problems helps me understand.

Name: ______________________________ Date: ______________

Unit 5 Practice 8: I can match expressions.

1) **Sample:** Draw a line to match the picture to the expression.

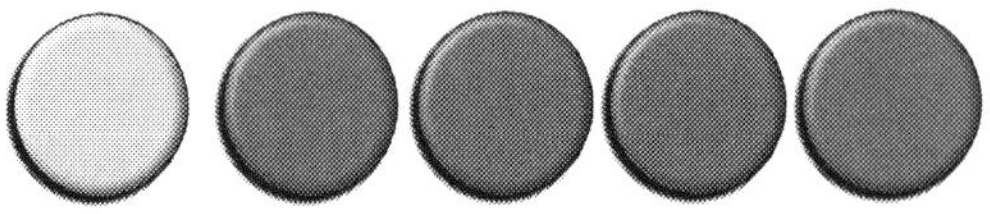

1 + 4	2 + 3

K.OA.3

2) Draw a line to match the picture to the expression.

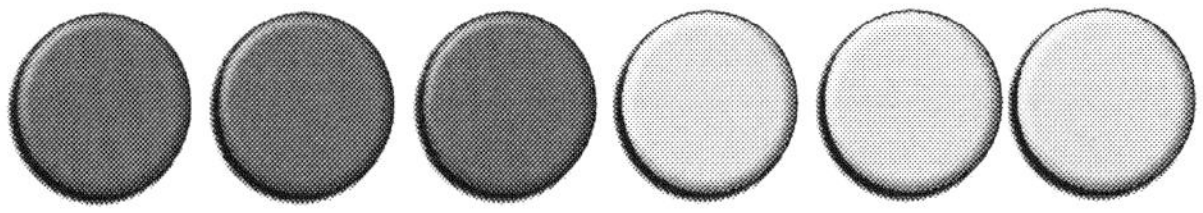

5 + 1	3 + 3

K.OA.3

Do you understand? **?** ✔

Affirmation: Acting out math problems helps me understand.

Name: ______________________________ Date: ______________

Unit 5 Practice 8: I can match expressions.

3) **Sample:** Draw a line to match the picture to the expression.

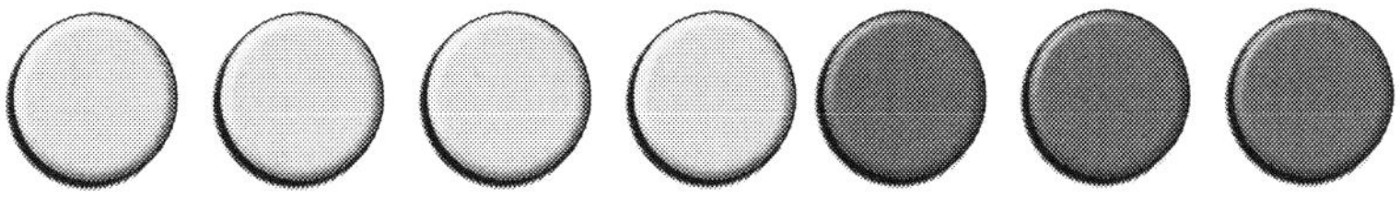

5 + 2   4 + 3

K.OA.3

4) Draw a line to match the picture to the expression.

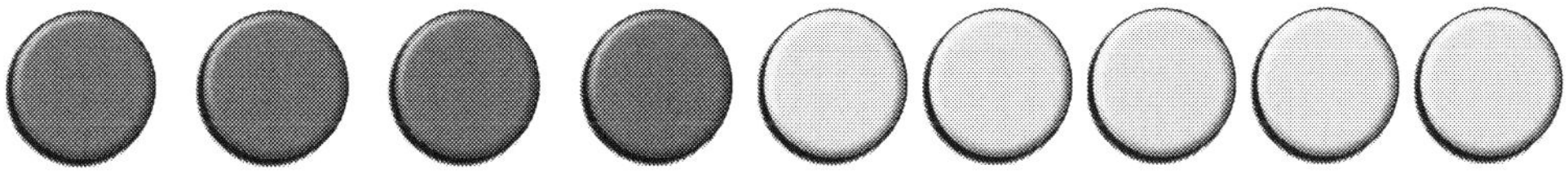

4 + 5   6 + 3

K.OA.3

Do you understand? **?** ✓

Name: ______________________________ Date: ______________

Unit 5 Practice 9: I can solve a story problem.

1) **Sample:** There were **5** ducks.
   3 ducks swim away.
   How many now?

5 - 3

There are 2 ducks left.

K.OA.2

2) There were **4** fish.
   2 fish swim up to meet them.
   How many now?

______ + ______

There are ______ fish now.

K.OA.2

Do you understand? **?** ✓

Name: ______________________ Date: __________

Unit 5 Practice 9: I can solve a story problem.

3) There were **7** pigs.
3 pigs walk away.
How many now?

____ − ____

There are ______ pig now.

K.OA.2

4) There were **3** ants.
2 ants join the fun.
How many now?

____ + ____

There are ______ ants in all.

K.OA.2

Do you understand? **?** ✓

Affirmation: I explain and justify my thinking.

Name: ______________________________ Date: ______________

Unit 5 Practice 10: I can match a 10-frame to a number.

1) **Sample:** Draw a line to match the 10-frame to the number.

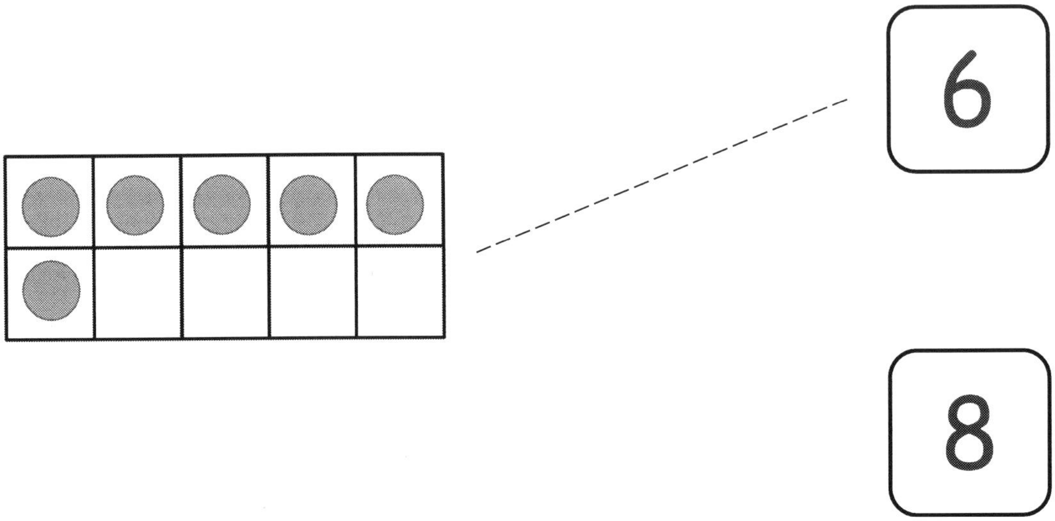

K.CC.4b

2) Draw a line to match the 10-frame to the number.

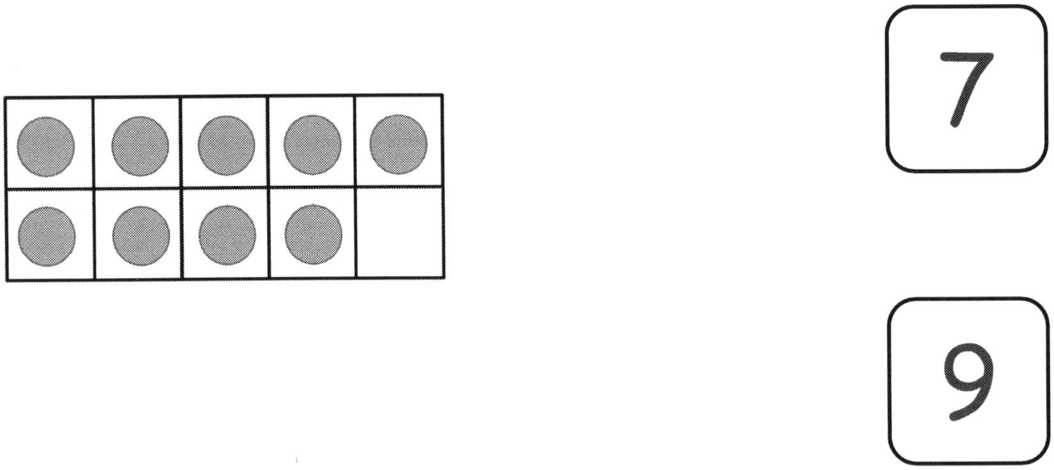

K.CC.4b

Do you understand? **?** ✓

Name: ______________________________ Date: ______________

Unit 5 Practice 10: I can match a 10-frame to a number.

3) Draw a line to match the 10-frame to the number.

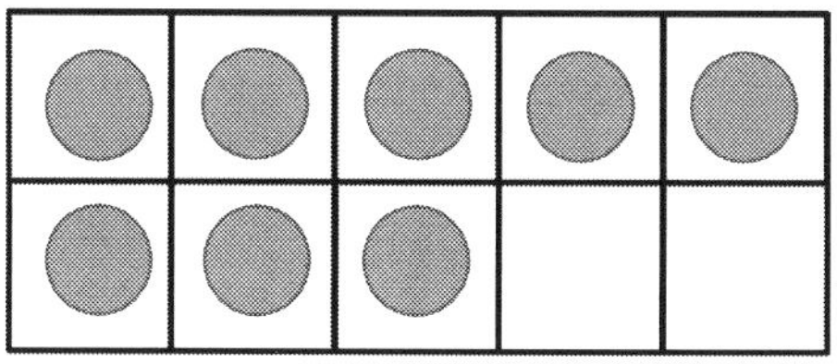

7

8

K.CC.4b

4) Draw a line to match the 10-frame to the number.

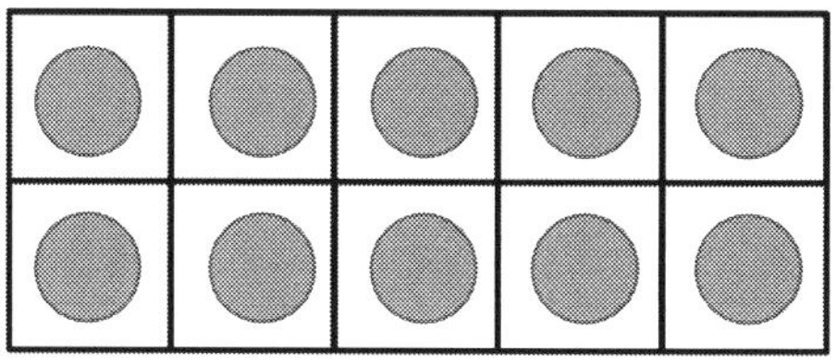

9

10

K.CC.4b

Do you understand? ? ✓

Affirmation: I explain and justify my thinking.

Name: ______________________________ Date: ______________

Unit 5 Practice 11: I can match a 10-frame to an equation.

1) **Sample:** Draw a line to match the 10-frame to the equation.

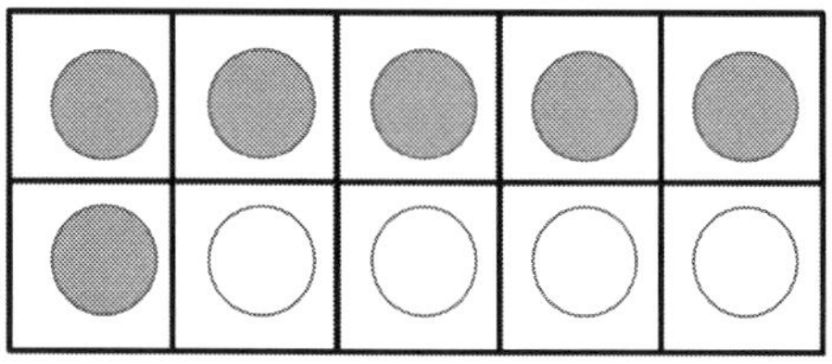

10 = 6 + 4

10 = 7 + 3

K.OA.3

2) Draw a line to match the 10-frame to the equation.

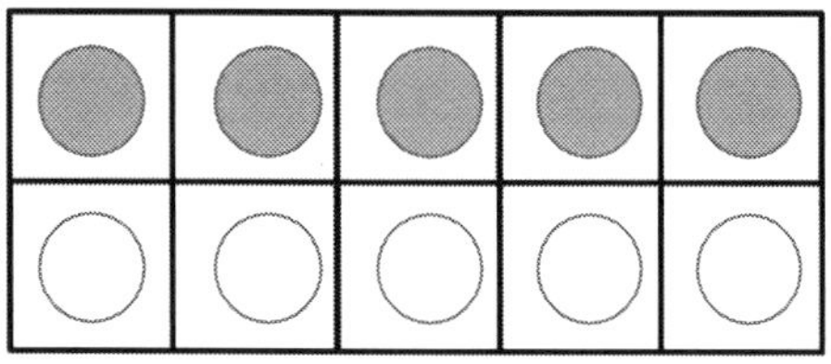

10 = 5 + 5

10 = 4 + 6

K.OA.3

Do you understand? ? ✓

Affirmation: I explain and justify my thinking.

Name: ______________________________ Date: ______________

Unit 5 Practice 11: I can match a 10-frame to an equation.

3) Draw a line to match the 10-frame to the equation.

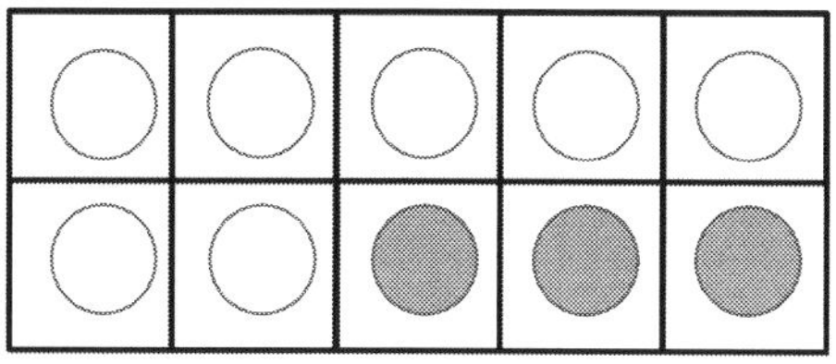

10 = 5 + 5

10 = 7 + 3

K.OA.3

4) Draw a line to match the 10-frame to the equation.

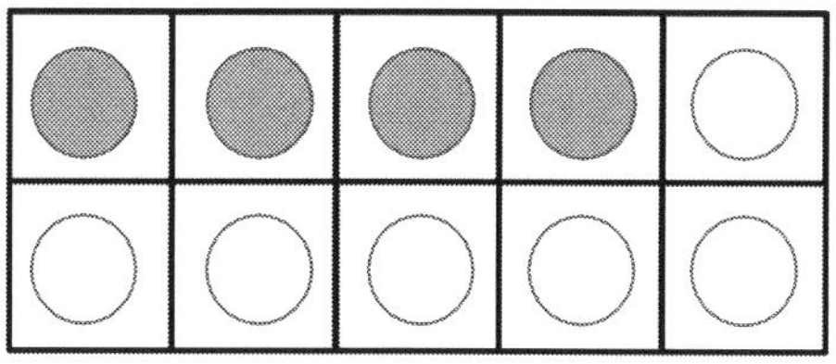

10 = 4 + 6

10 = 3 + 7

K.OA.3

Do you understand? ? ✓

Affirmation: I explain and justify my thinking.

Name: ______________________________ Date: ______________

Unit 5 Practice 12: I can create a 10-frame to match an equation.

1) **Sample:** Draw and color counters in 10-frame to match the equation.

$$10 = 7 + 3$$

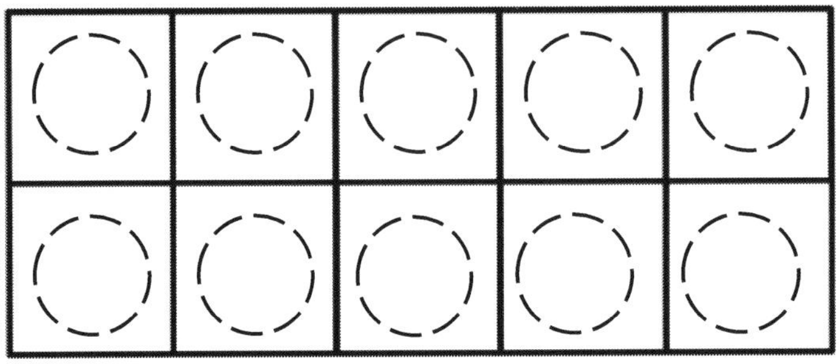

K.OA.3

2) Draw and color counters in the 10-frame to match the equation.

$$10 = 6 + 4$$

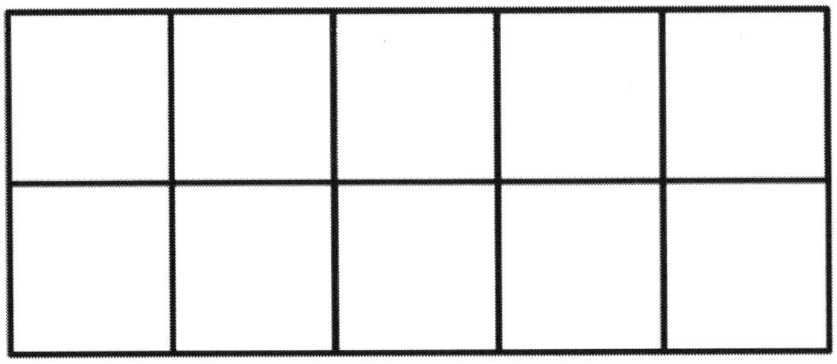

K.OA.3

Do you understand? **?** ✓

Affirmation: I explain and justify my thinking.

Name: ______________________________ Date: ______________

Unit 5 Practice 12: I can create a 10-frame to match an equation.

3) Draw and color counters in the 10-frame to match the equation.

$$10 = 5 + 5$$

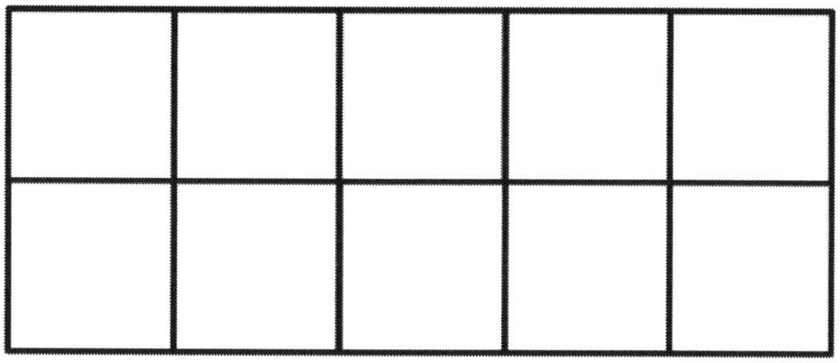

K.OA.3

4) Draw and color counters in the 10-frame to match the equation.

$$10 = 8 + 2$$

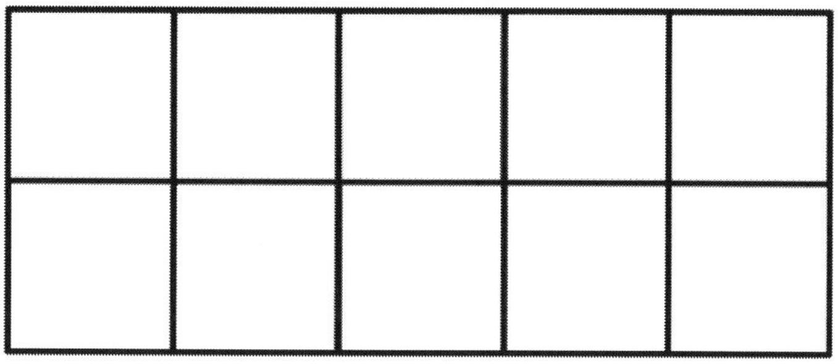

K.OA.3

Do you understand? ? ✓

Name: ______________________________ Date: ______________

Unit 5 Practice 13: I can make 10.

1) **Sample:** Complete the equation.
Draw counters in the 10-frame to make the equation true.

$$10 = 6 + \underline{4}$$

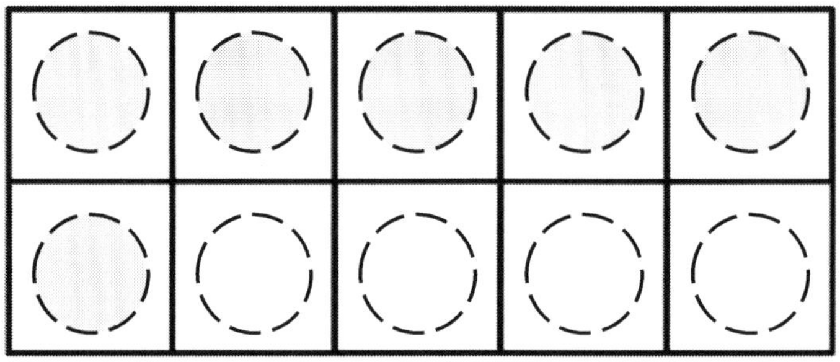

K.OA.3

2) Complete the equation.
Draw counters in the 10-frame to make the equation true.

$$10 = 7 + \_\_\_\_$$

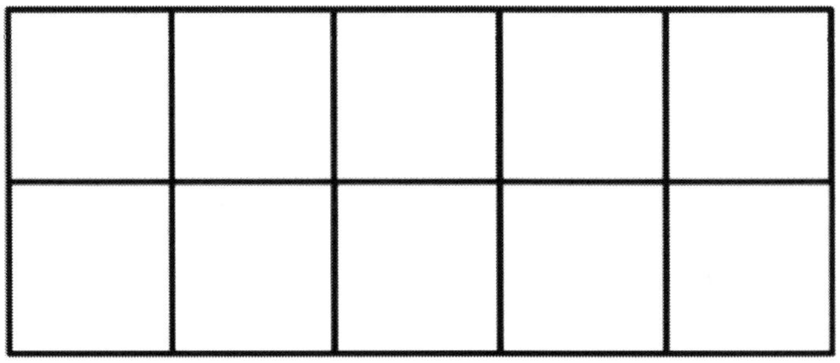

K.OA.3

Do you understand? **?** ✓

Affirmation: I explain and justify my thinking.

Name: ____________________ Date: ____________

Unit 5 Practice 13: I can make 10.

3) Complete the equation.
Draw counters in the 10-frame to make the equation true.

$$10 = 5 + \_\_\_\_$$

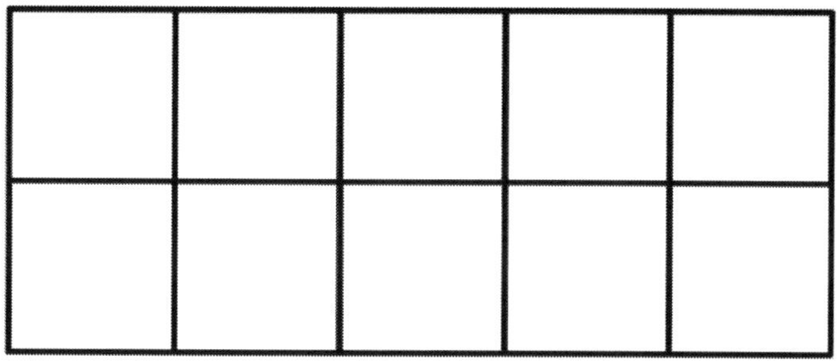

K.OA.3

4) Complete the equation.
Draw counters in the 10-frame to make the equation true.

$$10 = 8 + \_\_\_\_$$

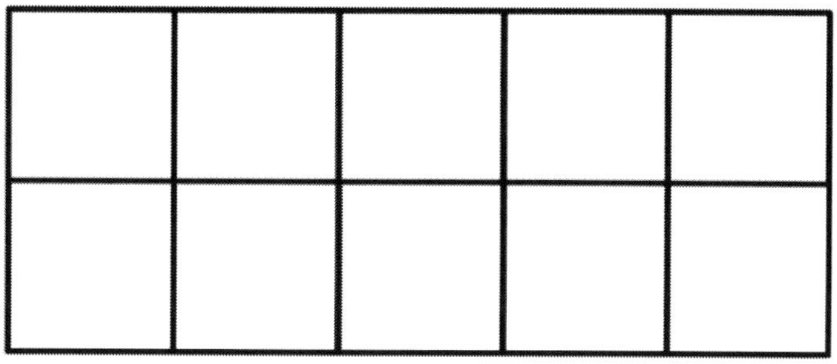

K.OA.3

Do you understand? **?** ✓

Name: ______________________________ Date: ____________

Unit 5 Practice 14: I can make 10.

1) **Sample:** Use the 10-frame to complete the equation.

$$10 = \underline{7} + \underline{3}$$

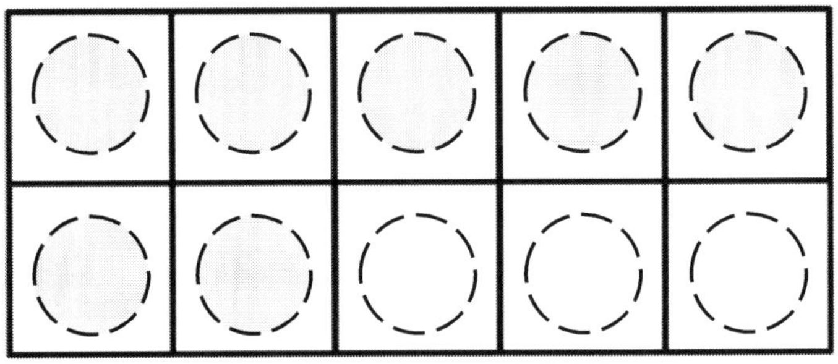

K.OA.3

2) Use the 10-frame to complete the equation.

$$10 = \underline{\quad\quad} + \underline{\quad\quad}$$

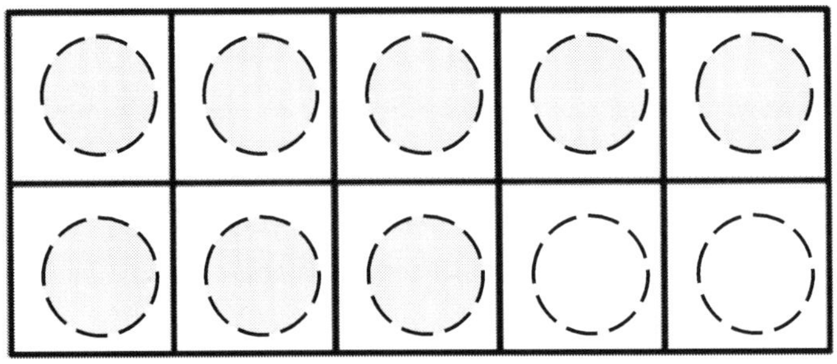

K.OA.3

Do you understand? **?** ✓

Name: ______________________________ Date: ______________

Unit 5 Practice 14: I can make 10.

3) Use the 10-frame to complete the equation.

$$10 = \_\_\_\_ + \_\_\_\_$$

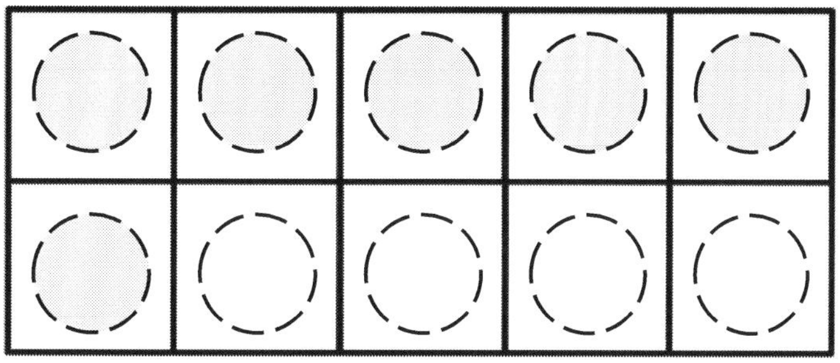

K.OA.3

4) Use the 10-frame to complete the equation.

$$10 = \_\_\_\_ + \_\_\_\_$$

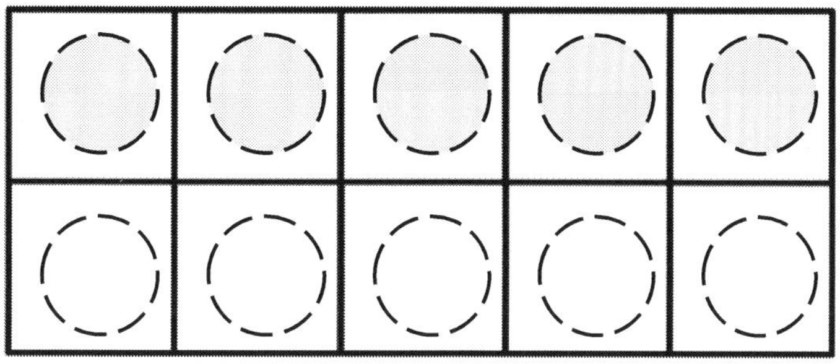

K.OA.3

Do you understand? ? ✓

Affirmation: I am a problem solver!

Name: ______________________________ Date: ______________

Unit 5 Practice 15: I can solve a story problem.

1) **Sample:** Carlos had **4** pencils.
Gina gives Carlos **3** more pencils.
How many pencils does Carlos have now?
Draw a picture to solve the story problem.

Carlos has 7 pencils now.

K.OA.2

2) Helena had **3** crayons.
Jenny gives Helena **2** more crayons.
How many crayons does Helena have now?
Draw a picture to solve the story problem.

Helena has __________ crayons now.

K.OA.2

Do you understand? **?** ✓

Name: ______________________ Date: __________

Unit 5 Practice 15: I can solve a story problem.

3) David had **7** toy cars.
Jeff gives David **2** more toy cars.
How many toy cars does David have now?
Draw a picture to solve the story problem.

David has ________ toy cars now.

K.OA.2

4) Maria had **7** flowers.
Nancy gives Maria **3** more flowers.
How many flowers does Maria have now?
Draw a picture to solve the story problem.

Maria has ________ flowers now.

K.OA.2

Do you understand? **?** ✓

Name: ____________________ Date: __________

I can practice grade level fluencies.

Set 1: Color the counters to show **9** in different ways.

1) 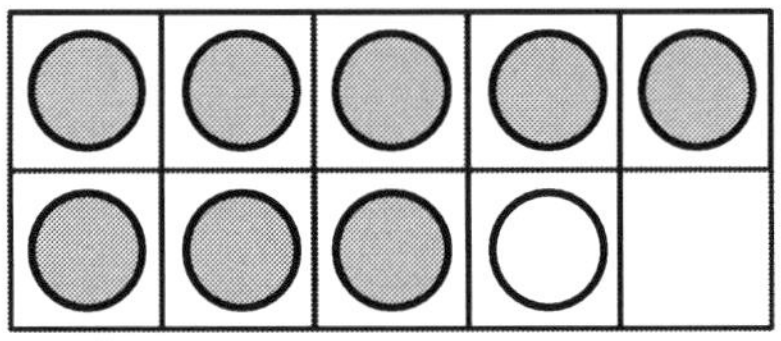

2) 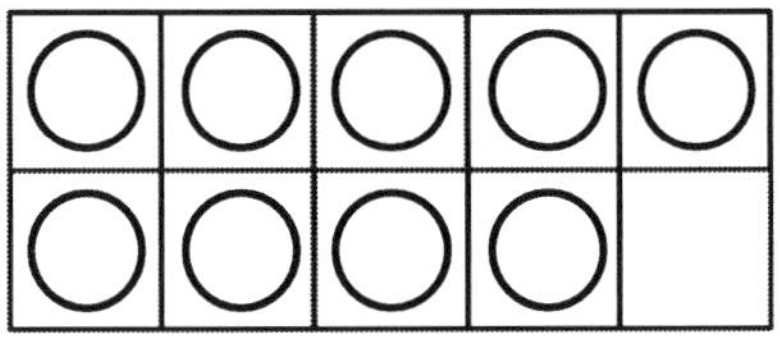

3) 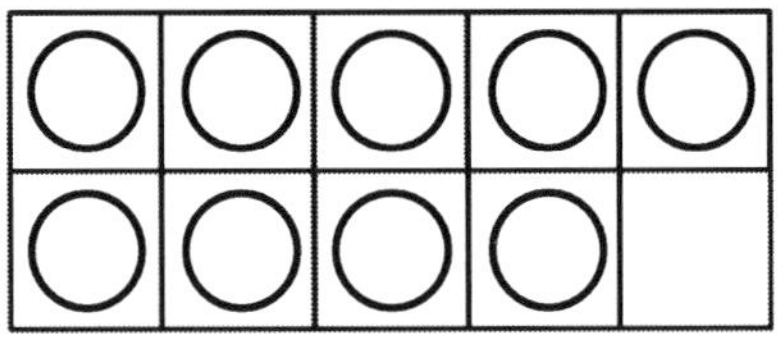

4) 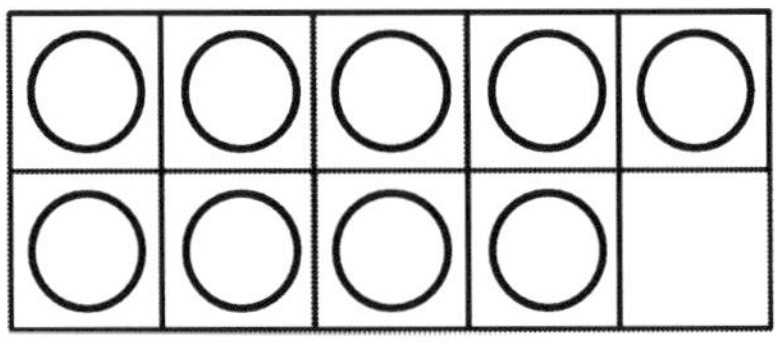

5) 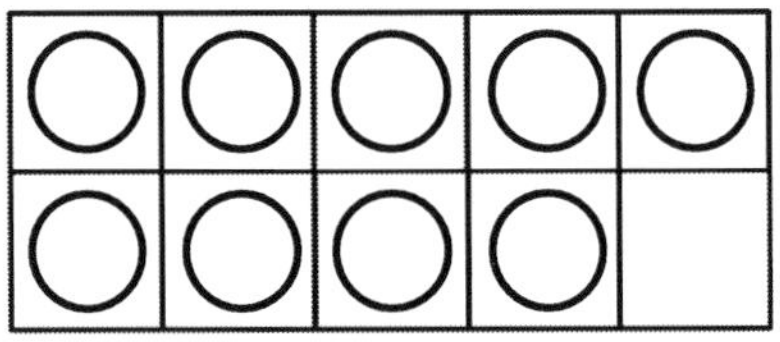

K.CC

Set 2: Write an expression to show the counters.

1) 

1 + 3

2) 

____ + ____

3) 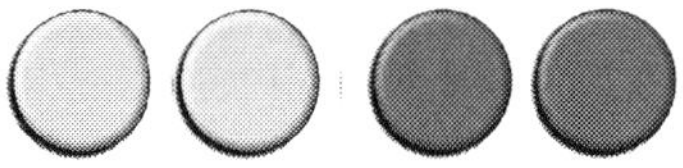

____ + ____

4) 

____ + ____

5) 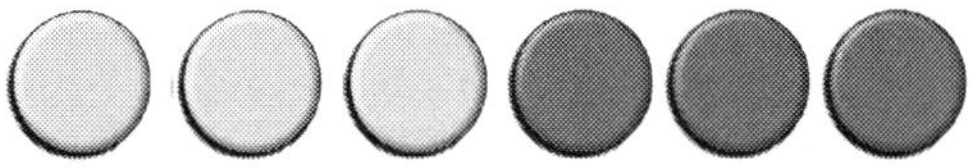

____ + ____

K.CC

Do you understand? **?** ✓

Unit Reflection

Composing and Decomposing Numbers to 10

Use this space to reflect on your understanding of the unit skills and concepts.

Skill/Concept	I can...	I need to work on...

Unit 6

Numbers 0–20

Numbers 0-20 is all about counting groups of objects within 20. Students also show groups of objects within 20 and explore tens and ones

Use a 10-Frame

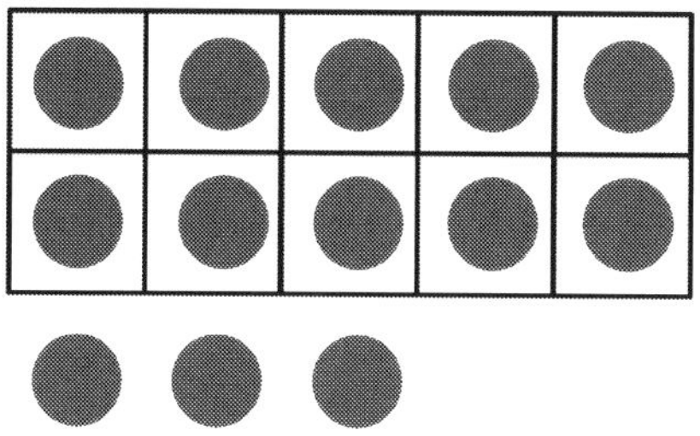

10 and 3 more is 13

Use Cubes

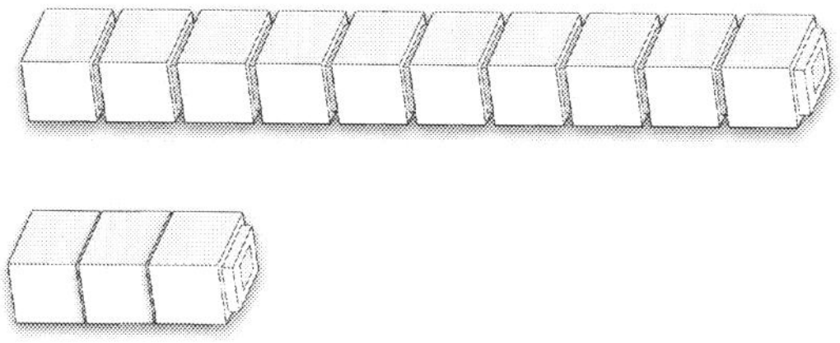

10 and 3 more is 13

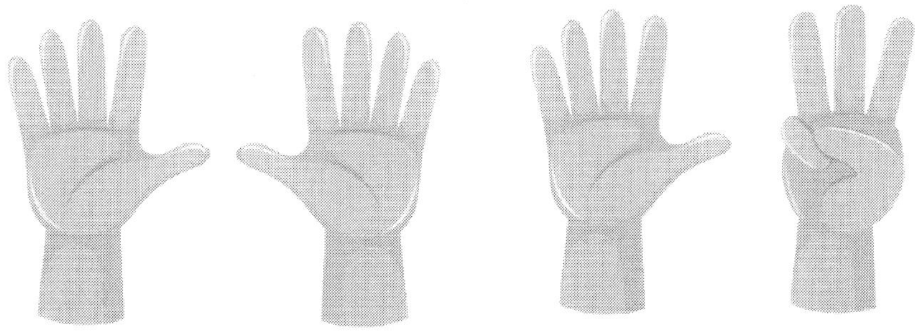

18 Fingers

20 Dots

Unit Vocabulary

Numbers 0 - 20

Use this space to visualize the math vocabulary for this unit.

Word or Phrase	Visual Reminder

Unit Tools

Numbers 0 - 20

Use this space to visualize the tools for this unit.

Tools	This is a...	It is used to...

Name: ______________________________ Date: ______________

Unit 6 Practice 1: I can count objects in a collection up to 20.

1) **Sample:** Color the cubes to show the equation:

$$14 = 10 + 4$$

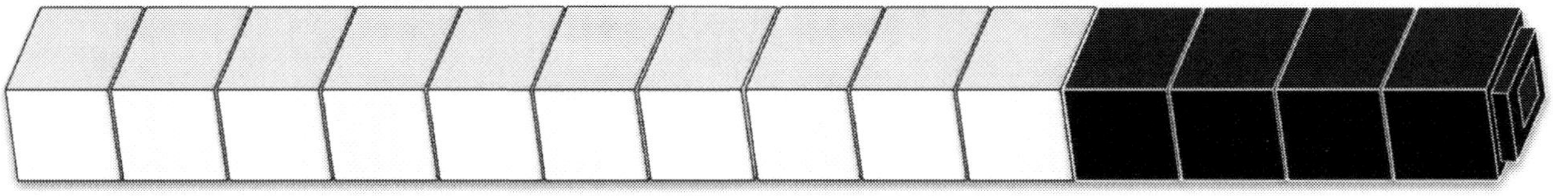

K.OA.1

2) Color the cubes to show the equation:

$$15 = 10 + 5$$

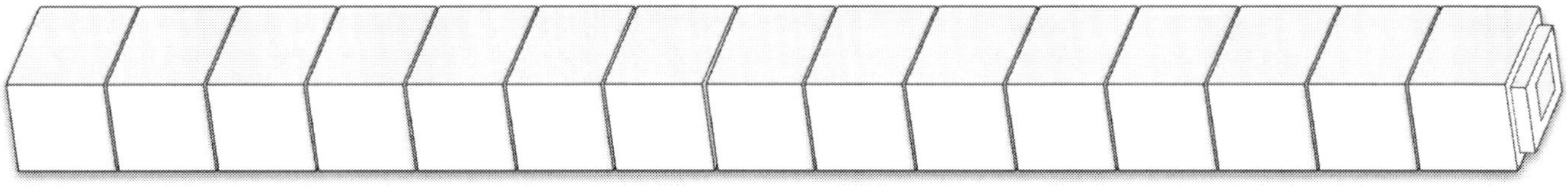

K.OA.1

Do you understand? **?** ✓

Name: ______________________ Date: __________

Unit 6 Practice 1: I can count objects in a collection up to 20.

3) Color the cubes to show the equation:

$$17 = 10 + 7$$

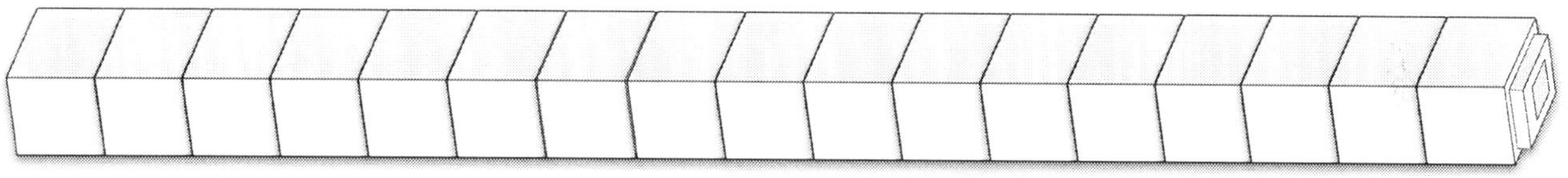

K.OA.1

4) Color the cubes to show the equation:

$$13 = 10 + 3$$

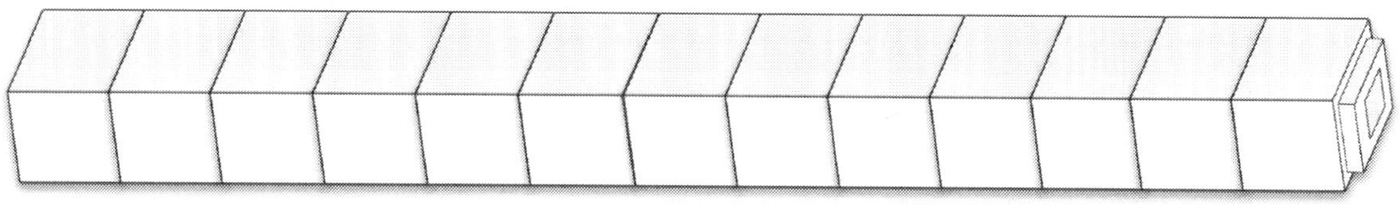

K.OA.1

Do you understand? ? ✓

Name: ______________________ Date: ____________

Unit 6 Practice 2: I can count objects in a collection up to 20.

1) **Sample:** Show 13 ones using one 10-frame.

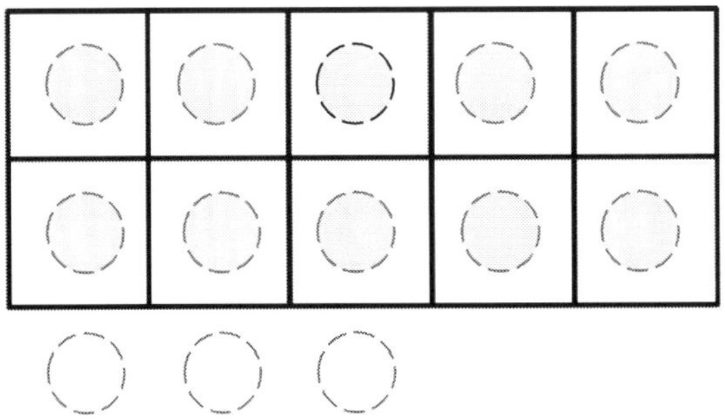

There are 10 ones and <u>3</u> more ones.

K.NBT.1

2) Show 12 ones using one 10-frame.

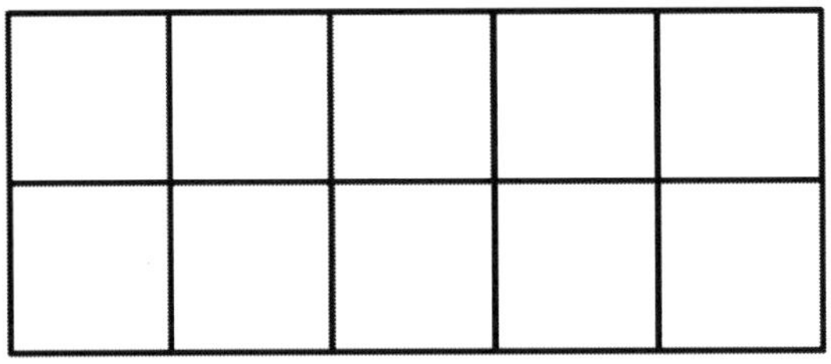

There are 10 ones and ________ more ones.

K.NBT.1

Do you understand? ? ✓

Affirmation: I am a problem solver!

Name: ____________________ Date: __________

Unit 6 Practice 2: I can count objects in a collection up to 20.

3) Show 15 ones using one 10-frame.

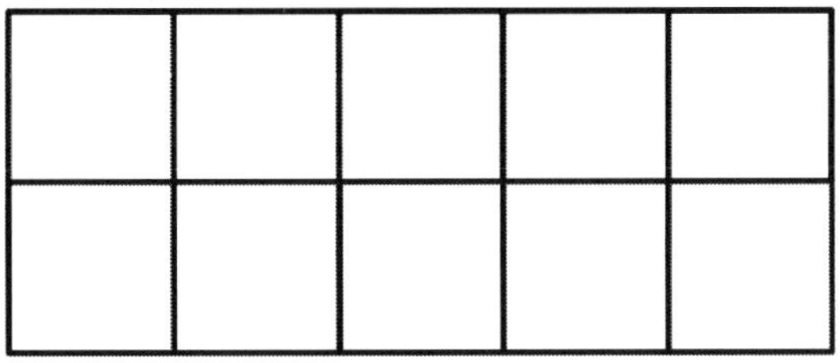

There are 10 ones and ________ more ones.

K.NBT.1

4) Show 14 ones using one 10-frame.

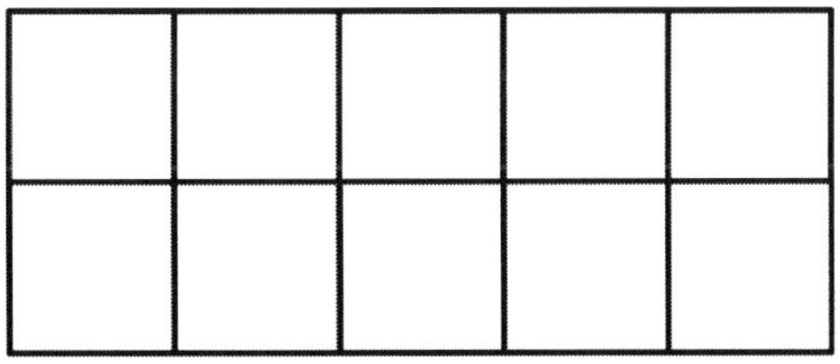

There are 10 ones and ________ more ones.

K.NBT.1

Do you understand? **?** ✓

Name: ________________________________ Date: ______________

Unit 6 Practice 3: I can count objects in a collection up to 20.

1) **Sample:** How many cubes?

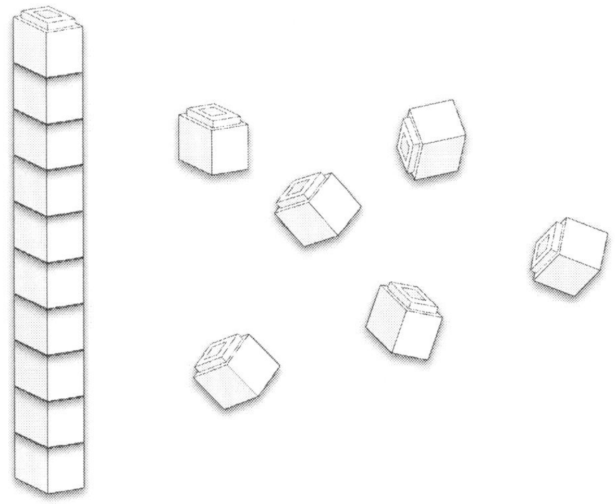

There are **16** cubes.

K.NBT.1

2) How many cubes?

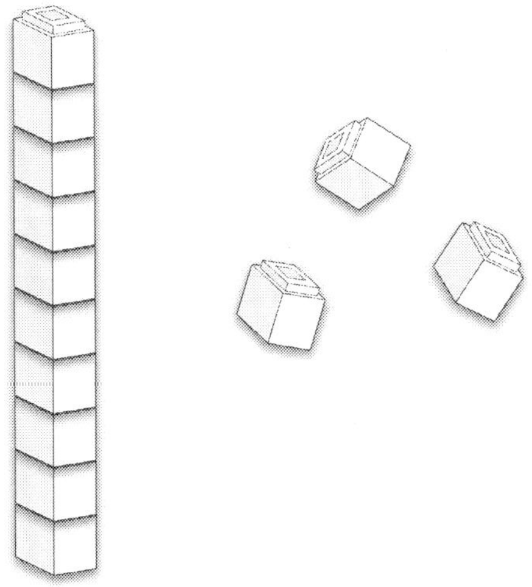

There are __________ cubes.

K.NBT.1

Do you understand? ? ✓

Affirmation: I am a problem solver!

Name: ______________________ Date: ____________

Unit 6 Practice 3: I can count objects in a collection up to 20.

3) How many cubes?

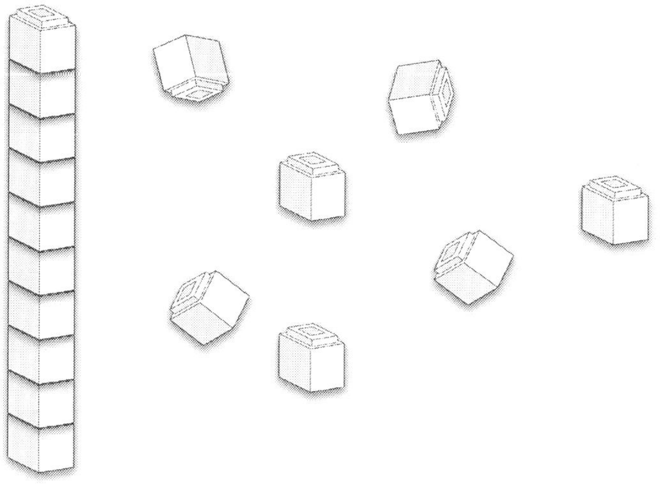

There are ________ cubes.

K.NBT.1

4) How many cubes?

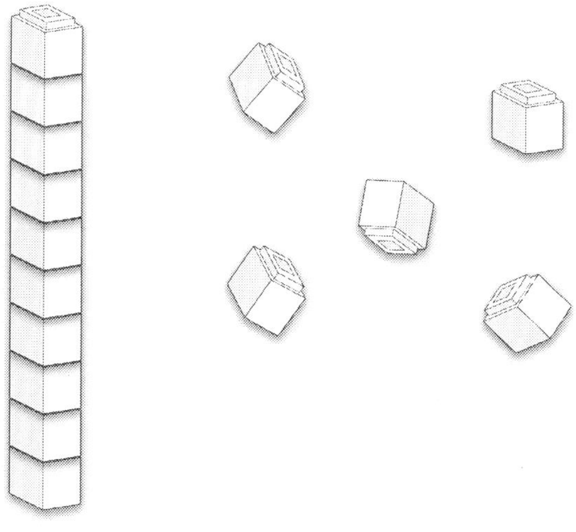

There are ________ cubes.

K.NBT.1

Do you understand? **?** ✔

Name: ______________________________ Date: ______________

Unit 6 Practice 4: I can count objects in a collection up to 20.

1) **Sample:** Does each group have the **same** amount?
Circle **Yes** or **No**.

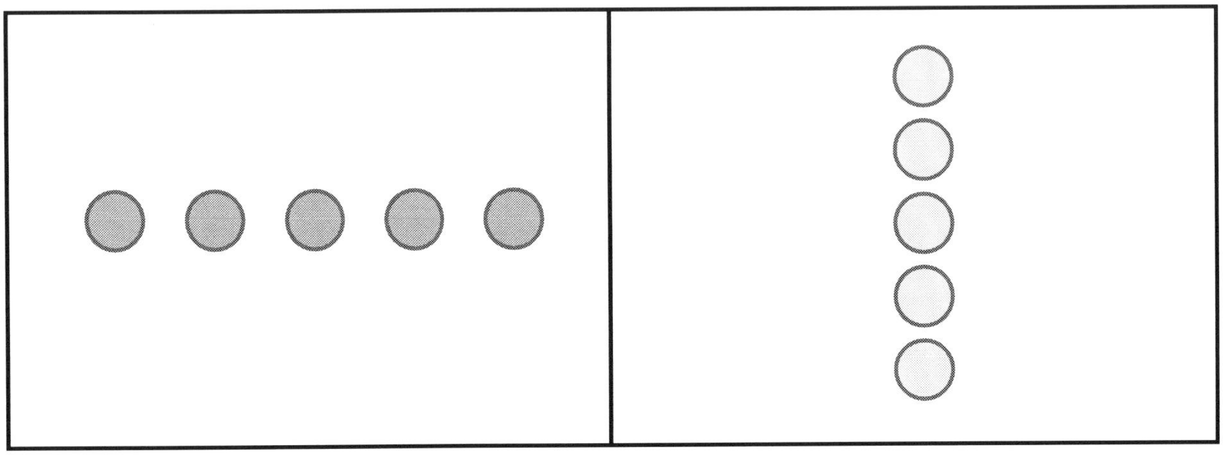

Yes **No**

K.CC.5

2) Does each group have the **same** amount?
Circle **Yes** or **No**.

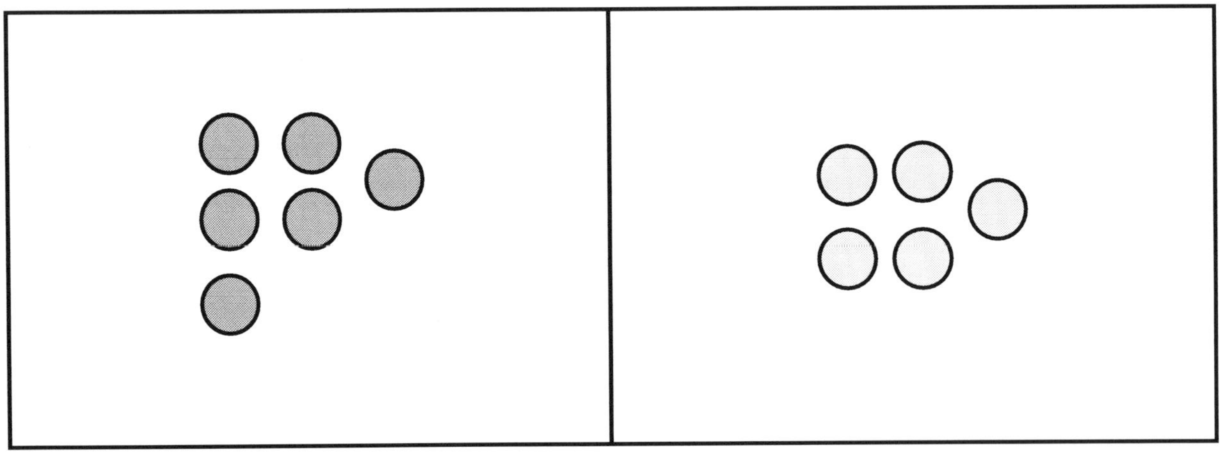

Yes **No**

K.CC.5

Do you understand? **?** ✔

Affirmation: I have a powerful mind!

Name: ______________________________ Date: ______________

Unit 6 Practice 4: I can count objects in a collection up to 20.

3) Does each group have the **same** amount?
Circle **Yes** or **No**.

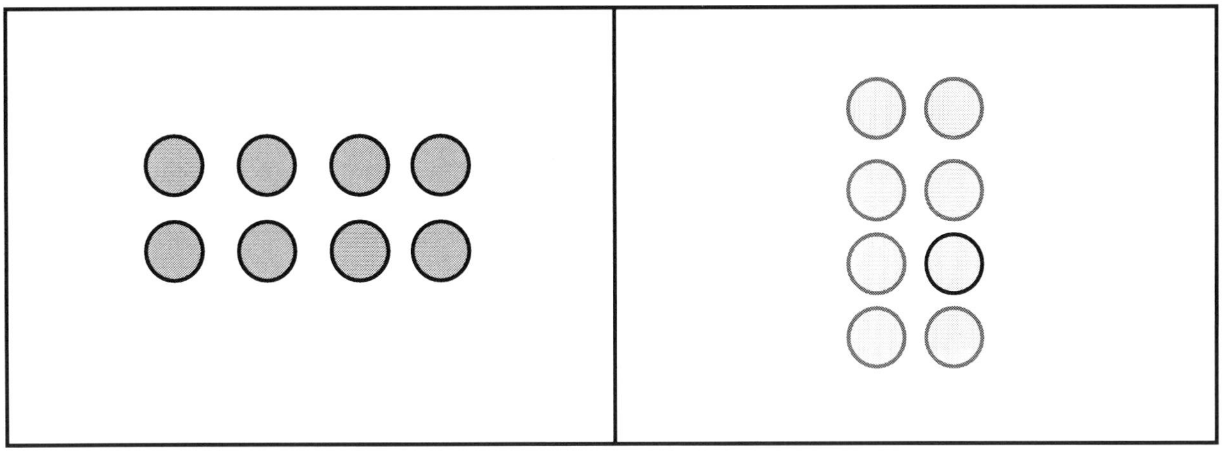

Yes **No**

K.CC.5

4) Does each group have the **same** amount?
Circle **Yes** or **No**.

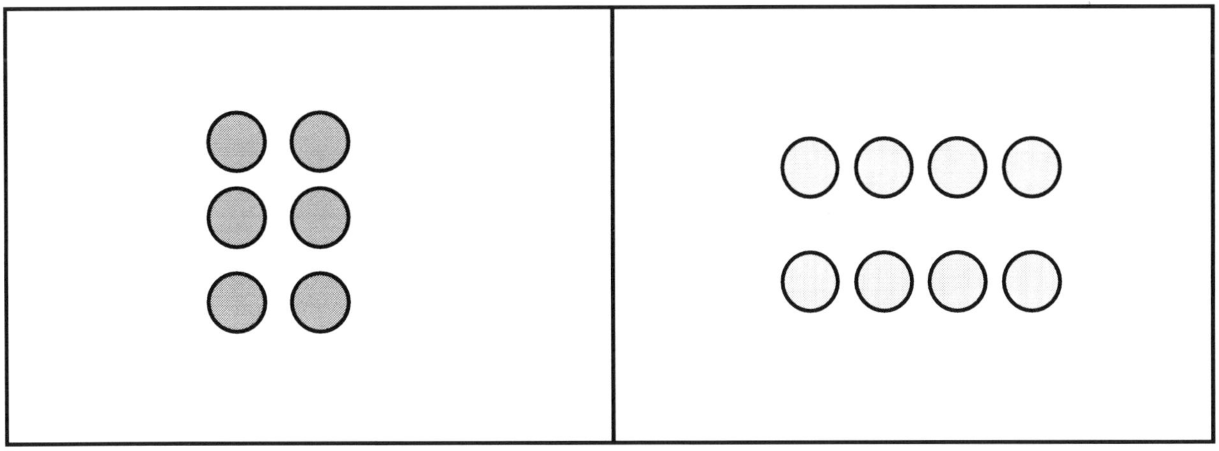

Yes **No**

K.CC.5

Do you understand? **?** ✓

Name: ______________________ Date: __________

Unit 6 Practice 5: I can count objects in a collection up to 20.

1) **Sample:** How many counters?
Circle the amount.

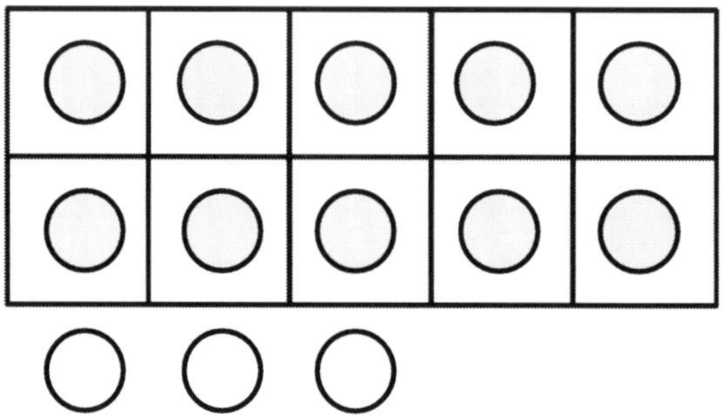

12 13 14

K.CC.5

2) How many counters?
Circle the amount.

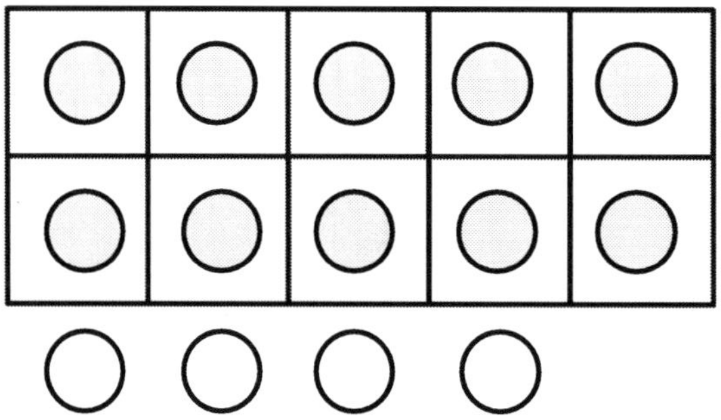

12 13 14

K.CC.5

Do you understand? ? ✓

Affirmation: I have a powerful mind!

Name: ______________________________ Date: ______________

Unit 6 Practice 5: I can count objects in a collection up to 20.

3) How many counters?
Circle the amount.

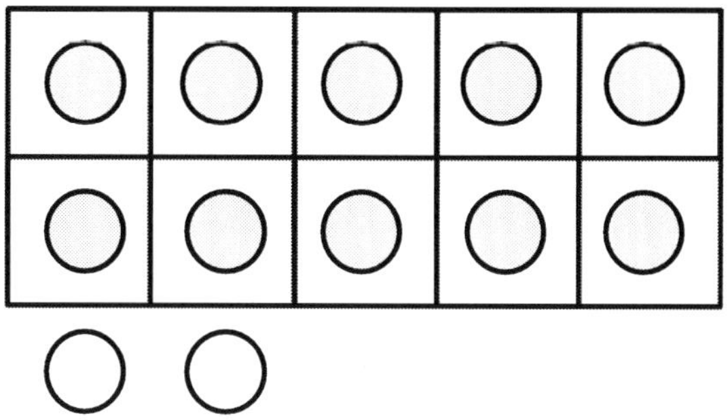

12 **13** **14**

K.CC,5

4) How many counters?
Circle the amount.

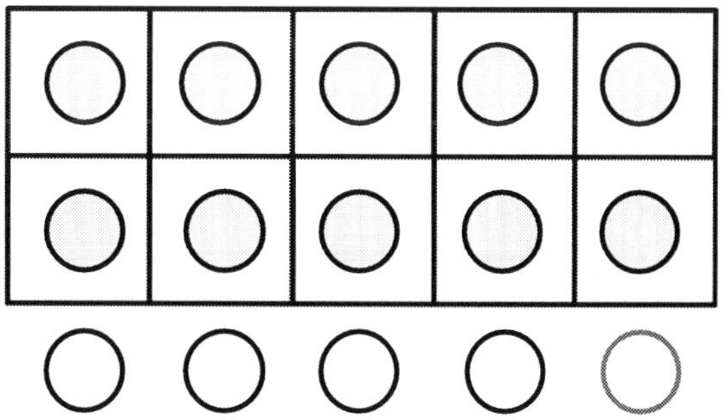

14 **15** **16**

K.CC.5

Do you understand? **?** ✓

Name: ______________________ Date: ____________

Unit 6 Practice 6: I can count objects in a collection up to 20.

1) **Sample:** How many fingers up?
Circle the amount.

12 13 14

K.CC.5

2) How many fingers up?
Circle the amount.

13 14 15

K.CC.5

Do you understand? ? ✓

Name: ____________________ Date: __________

Unit 6 Practice 6: I can count objects in a collection up to 20.

3) How many fingers up?
Circle the amount.

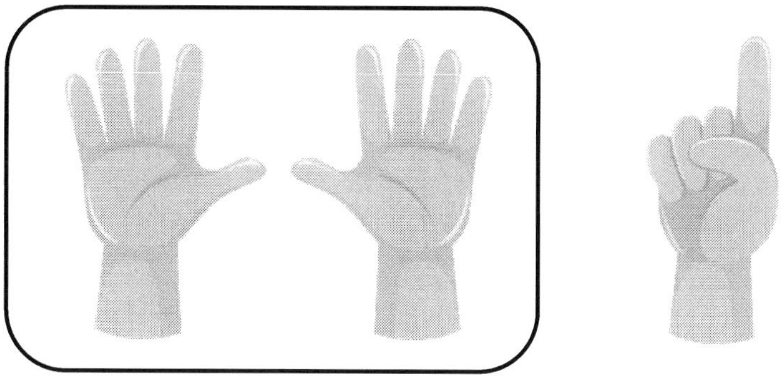

10 **11** **12**

K.CC.5

4) How many fingers up?
Circle the amount.

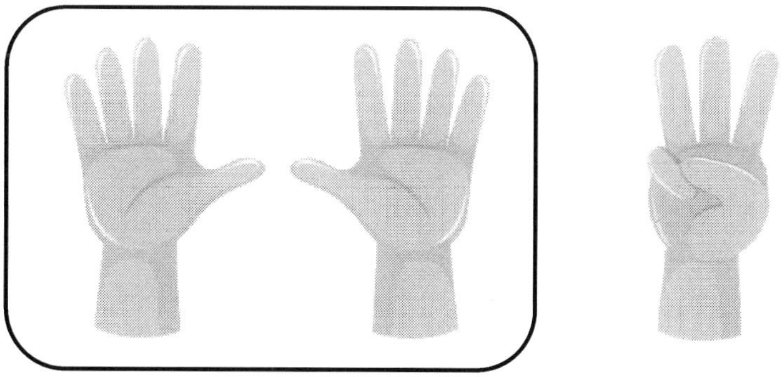

13 **14** **15**

K.CC.5

Do you understand? **?** ✓

Affirmation: I have a powerful mind!

Name: ______________________ Date: ____________

Unit 6 Practice 7: I can make numbers with a full 10-frame and more.

1) **Sample:** Draw **more** counters to show **14**. Write the amount.

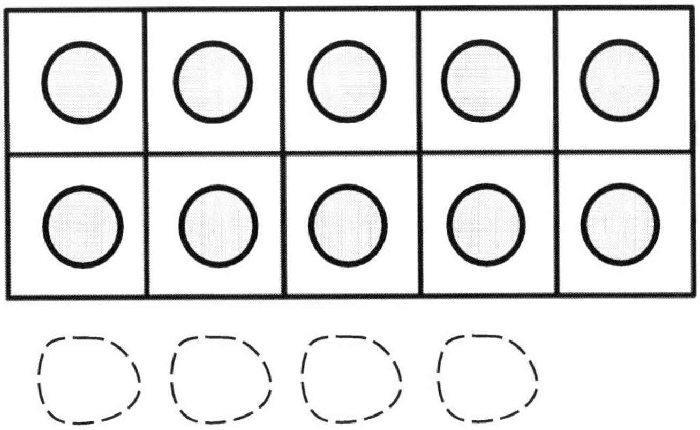

14 counters

K.OA.1

2) Draw **more** counters to show **12**. Write the amount.

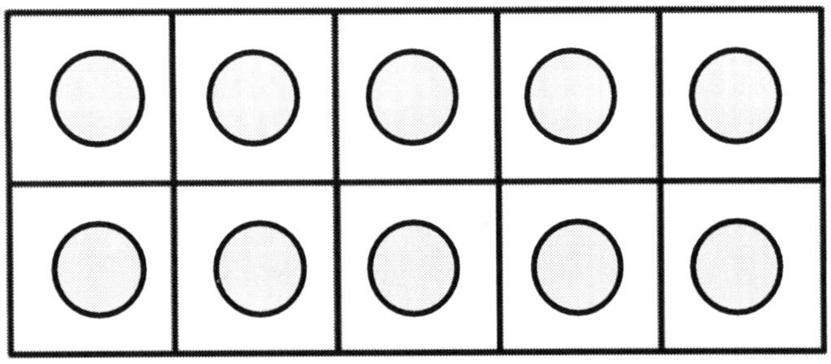

____ counters

K.OA.1

Do you understand? ? ✓

Affirmation: I have a powerful mind!

Name: ______________________________ Date: ______________

Unit 6 Practice 7: I can make numbers with a full 10-frame and more.

3) Draw **more** counters to show **15**. Write the amount.

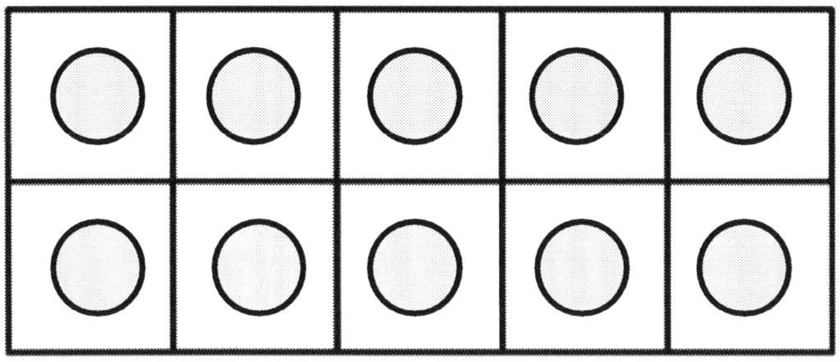

______ counters

K.OA.1

4) Draw **more** counters to show **13**. Write the amount.

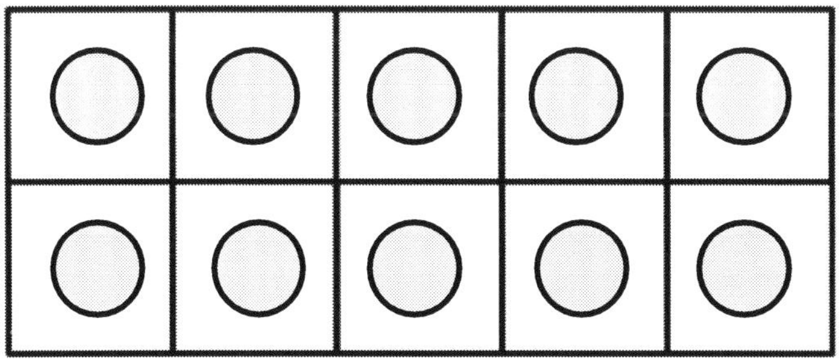

______ counters

K.OA.1

Do you understand? **?** ✓

Affirmation: I have a powerful mind!

Name: ____________________ Date: ____________

Unit 6 Practice 8: I can make numbers with a full 10-frame and more.

1) **Sample:** Draw more counters to show **13**. Complete the equation.

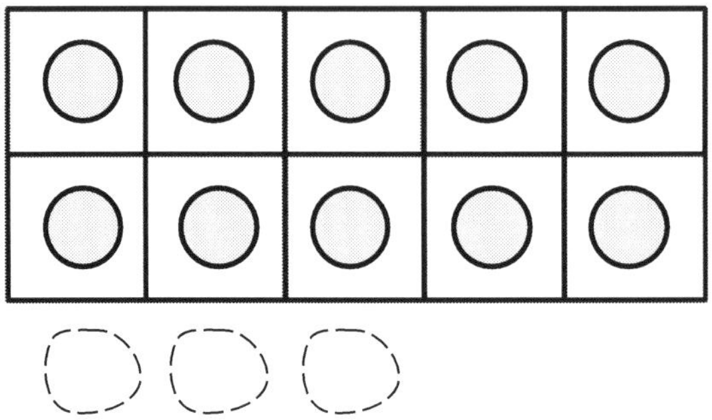

10 + ______ = 13

K.OA.1

2) Draw more counters to show **12**. Complete the equation.

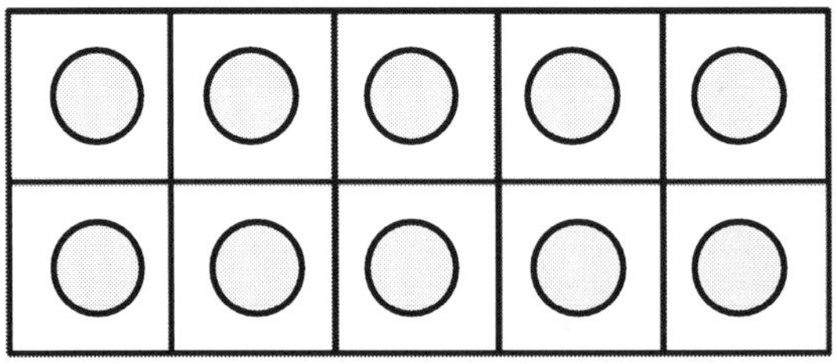

10 + ______ = 12

K.OA.1

Do you understand? **?** ✓

Name: ______________________________ Date: ______________

Unit 6 Practice 8: I can make numbers with a full 10-frame and more.

3) Draw more counters to show **15**. Complete the equation.

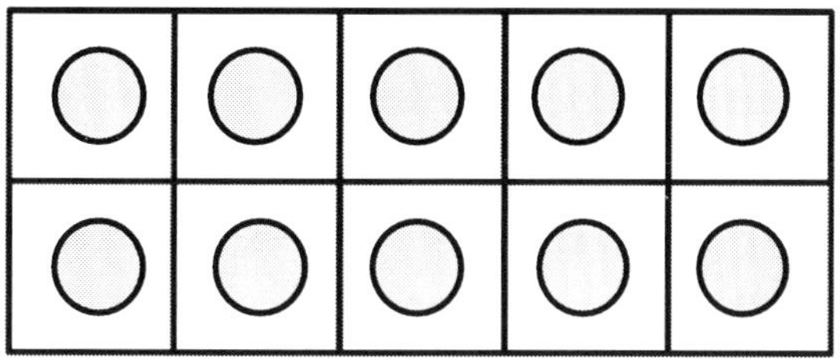

10 + ____ = 13

K.OA.1

4) Draw more counters to show **14**. Complete the equation.

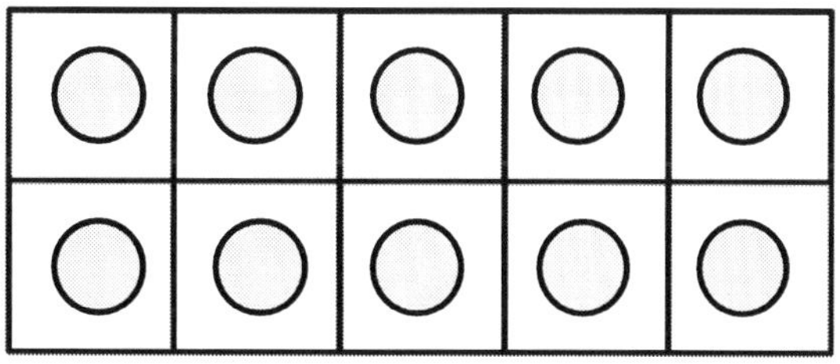

10 + ____ = 12

K.OA.1

Do you understand? **?** ✓

Affirmation: I am a mathematical thinker.

Name: ______________________________ Date: ______________

Unit 6 Practice 9: I can match 10-frames to equations.

1) **Sample:** Draw a line to match the 10-frame to the equation.

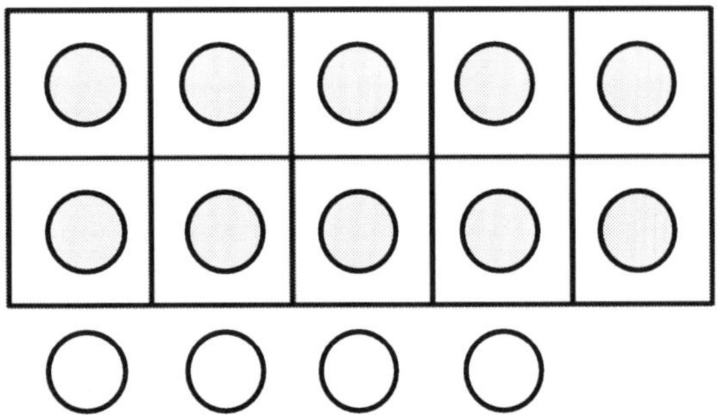

10 + 2 = 12

10 + 4 = 14

K.OA.1

2) Draw a line to match the 10-frame to the equation.

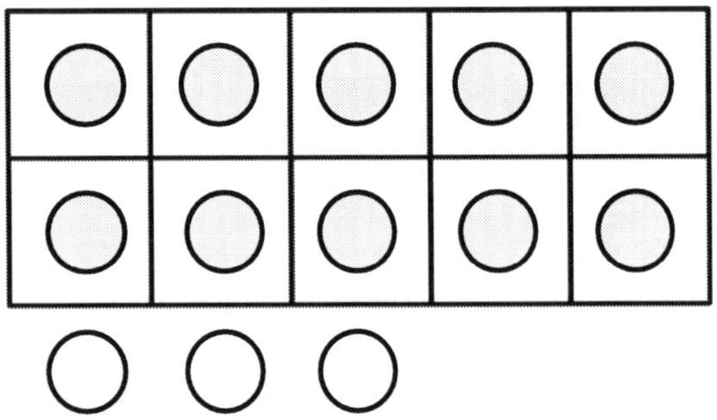

10 + 3 = 13

10 + 4 = 14

K.OA.1

Do you understand? **?** ✓

Name: ______________________________ Date: ______________

Unit 6 Practice 9: I can match 10-frames to equations.

3) Draw a line to match the 10-frame to the equation.

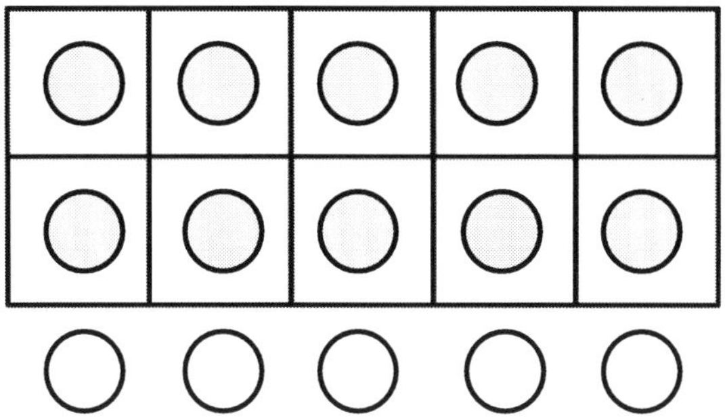

$10 + 5 = 15$

$10 + 6 = 16$

K.OA.1

4) Draw a line to match the 10-frame to the equation.

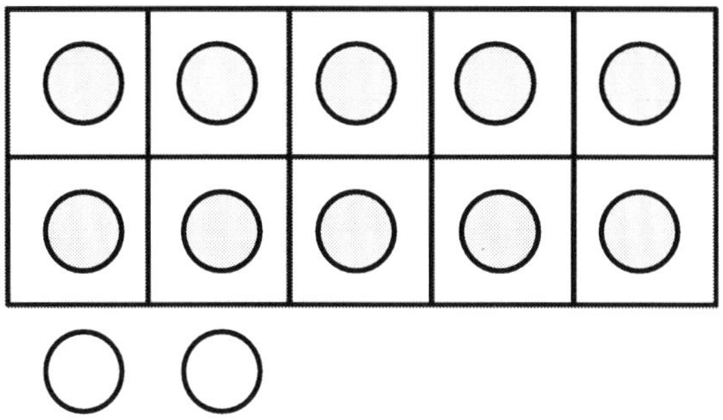

$10 + 3 = 13$

$10 + 2 = 12$

K.OA.1

Do you understand? **?** ✓

Name: ______________________________ Date: ______________

Unit 6 Practice 10: I can complete an equation.

1) **Sample:** Complete the equation to match the 10-frame.

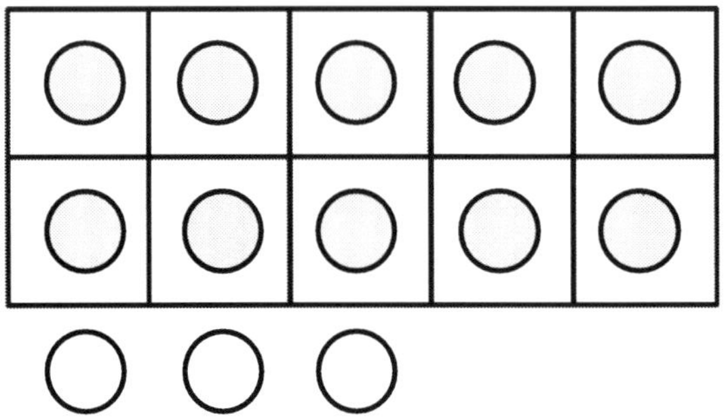

$10 + 3 =$ ______

K.OA.1

2) Complete the equation to match the 10-frame.

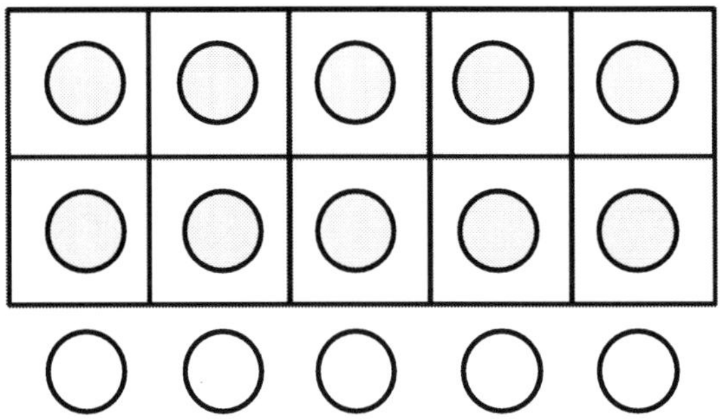

$10 + 5 =$ ______

K.OA.1

Do you understand? ? ✓

Name: ______________________________ Date: ______________

Unit 6 Practice 10: I can complete an equation.

3) Complete the equation to match the 10-frame.

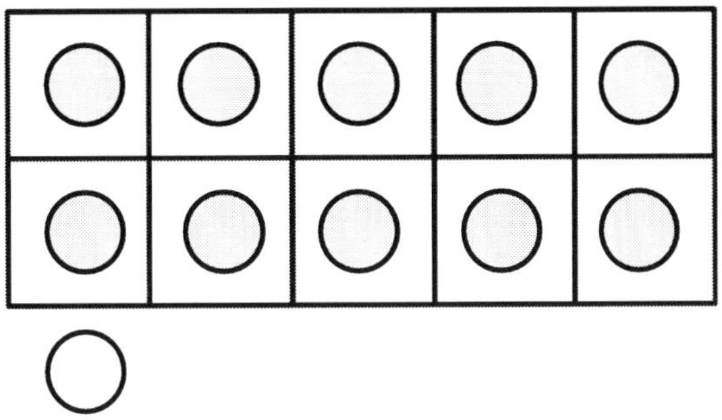

10 + 1 = ______

K.OA.1

4) Complete the equation to match the 10-frame.

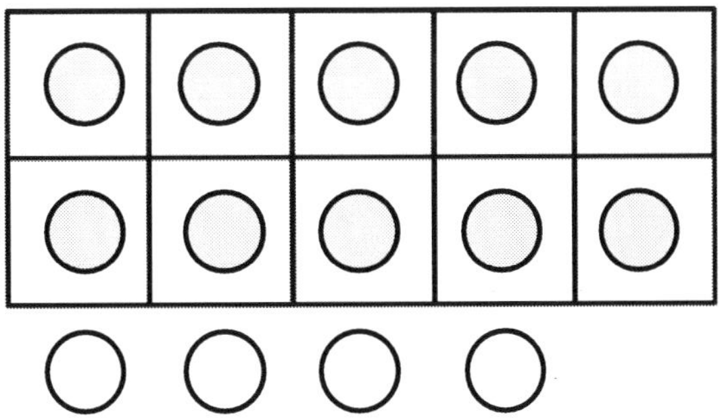

10 + 4 = ______

K.OA.1

Do you understand? ? ✓

Name: ______________________ Date: ____________

Unit 6 Practice 11: I can color counters to show an expression.

1) **Sample:** Color the counters to show the expression.

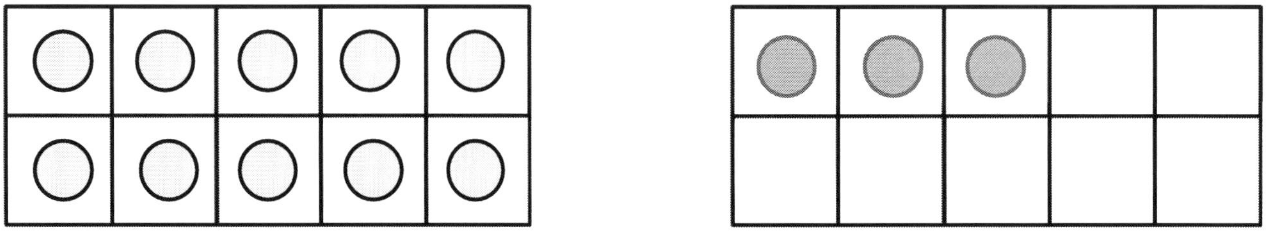

10 + 3

K.OA.1

2) Color the counters to show the expression.

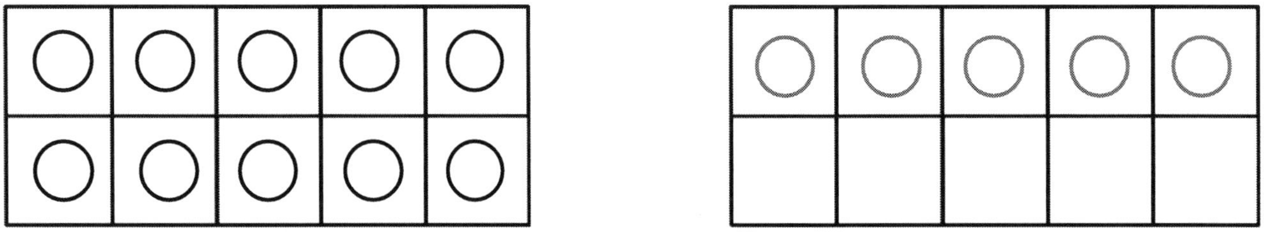

10 + 5

K.OA.1

Do you understand? **?** ✓

Name: ______________________ Date: ____________

Unit 6 Practice 11: I can color counters to show an expression.

3) Color the counters to show the expression.

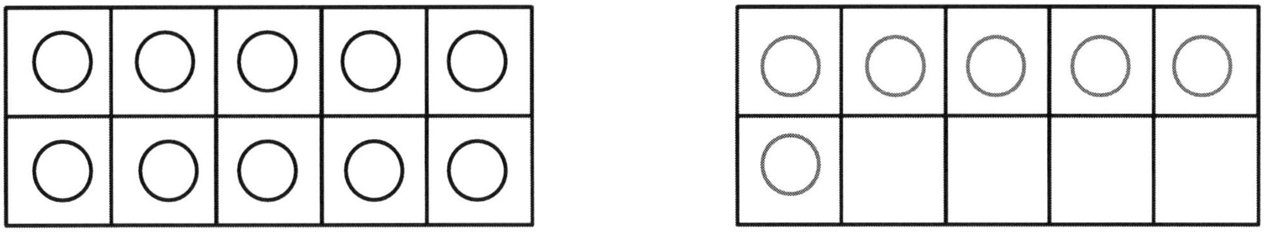

$$10 + 6$$

K.OA.1

4) Color the counters to show the expression.

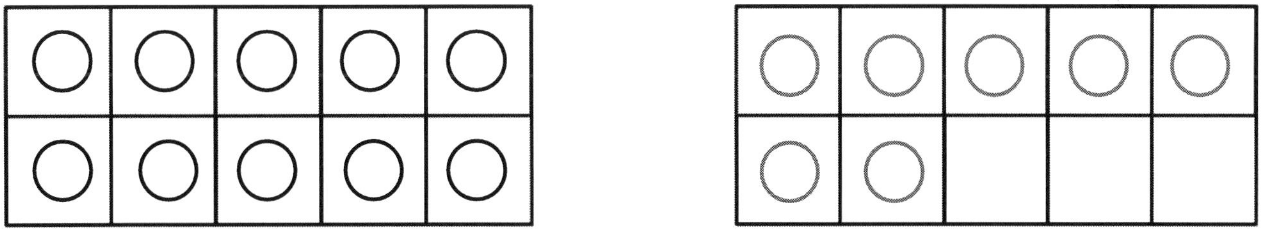

$$10 + 7$$

K.OA.1

Do you understand? **?** ✓

Affirmation: I am a mathematical thinker.

Name: ____________________ Date: ____________

Unit 6 Practice 12: I can count up to 20 objects.

1) **Sample:** Count the ducks. Write a number to show how many ducks.

There are __13__ ducks.

K.CC.5

2) Count the pigs. Write a number to show how many pigs.

There are ________ pigs.

K.CC.5

Do you understand? **?** ✓

Name: ______________________________ Date: ______________

Unit 6 Practice 12: I can count up to 20 objects.

3) Count the crayons. Write a number to show how many crayons.

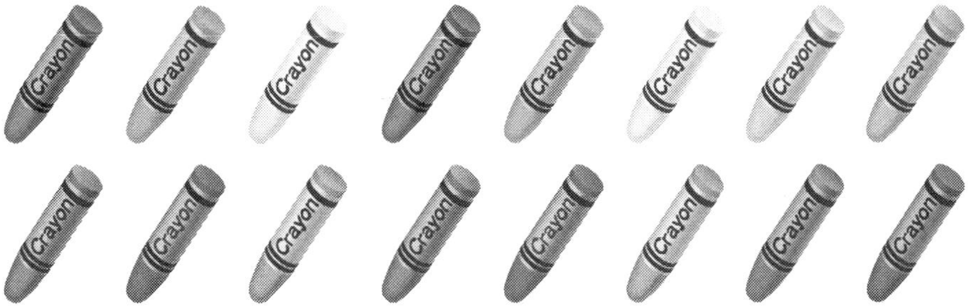

There are __________ crayons.

K.CC.5

4) Count the dogs. Write a number to show how many dogs.

There are __________ dogs.

K.CC.5

Do you understand? ? ✓

Name: ______________________________ Date: ______________

Unit 6 Practice 13: I can understand 10 ones and some more ones.

1) **Sample:** Show **13 ones** using one 10-frame. Complete the equation.

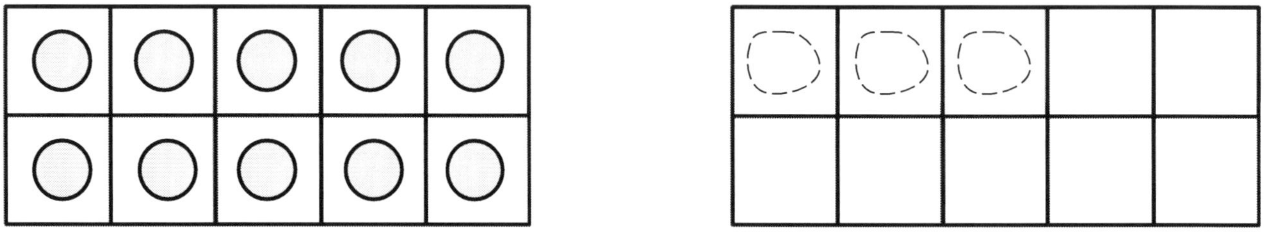

10 + 3 = 13

There are **10 ones** and 3 **more ones.**

K.NBT.1

2) Show **15 ones** using one 10-frame. Complete the equation.

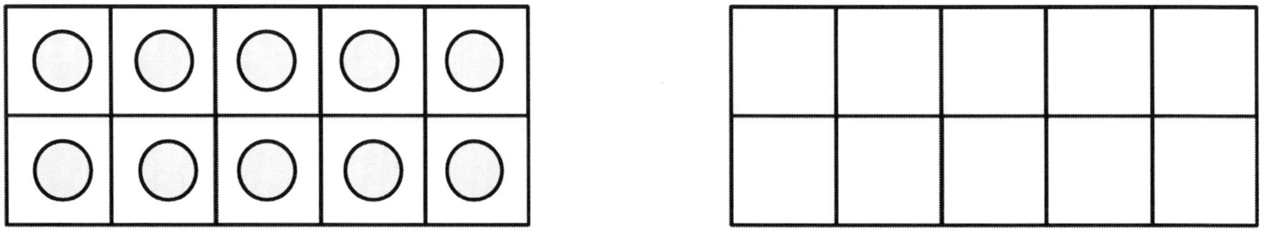

10 + ____ = 15

There are **10 ones** and ________ **more ones.**

K.NBT.1

Do you understand? **?** ✓

Affirmation: I am a mathematical thinker.

Name: ______________________________ Date: ______________

Unit 6 Practice 13: I can understand 10 ones and some more ones.

3) Show **16 ones** using one 10-frame. Complete the equation.

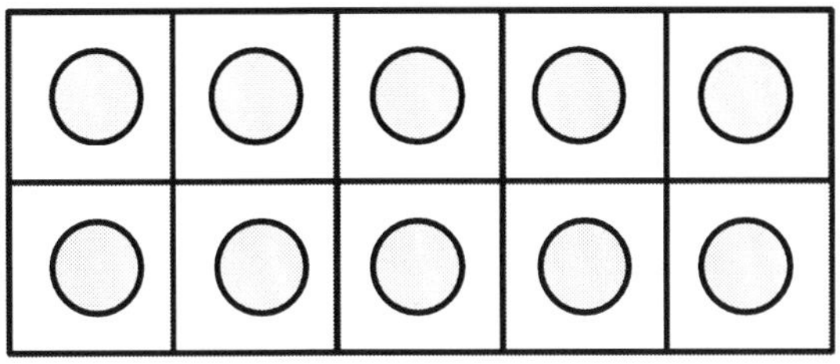

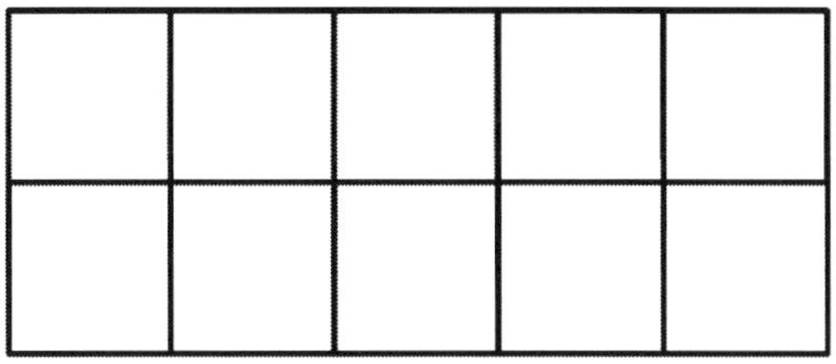

10 + ______ = 16

There are **10 ones** and ________ **more ones.**

K.NBT.1

4) Show **17 ones** using one 10-frame. Complete the equation.

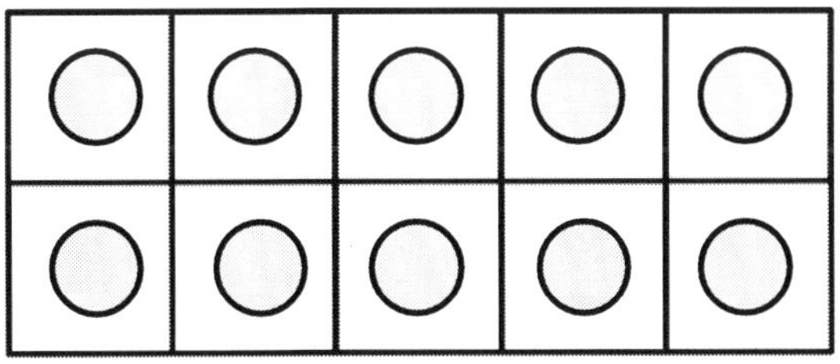

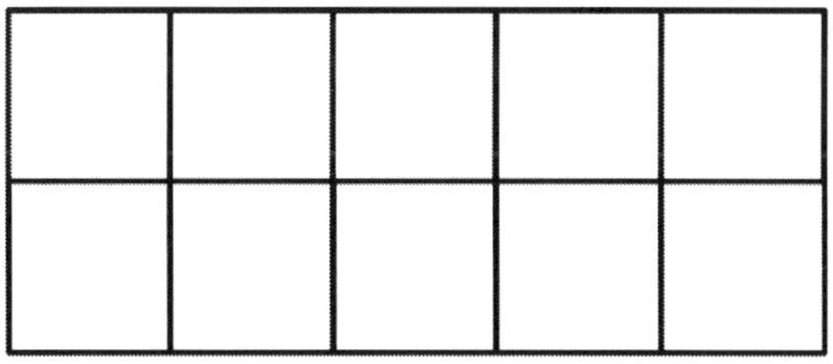

10 + ______ = 17

There are **10 ones** and ________ **more ones.**

K.NBT.1

Do you understand? **?** ✓

Name: ______________________________ Date: ______________

I can practice grade level fluencies.

Set 1: Match the counters to the number.

1) 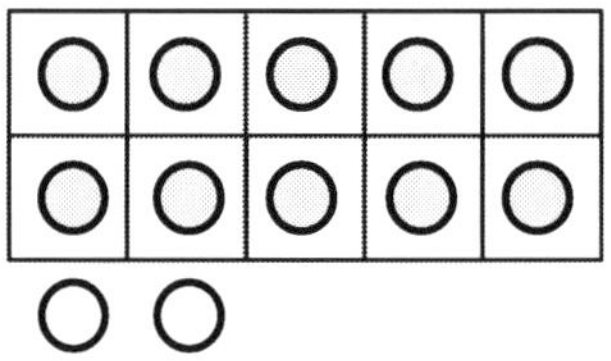

2) 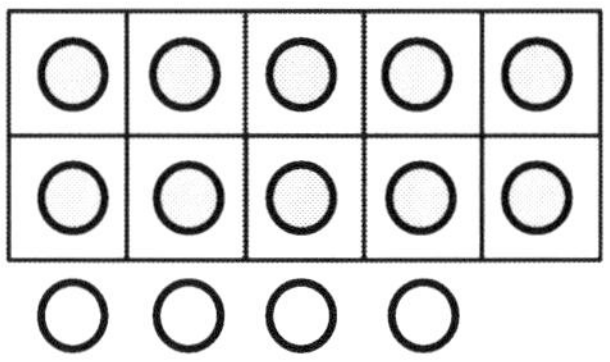

3) 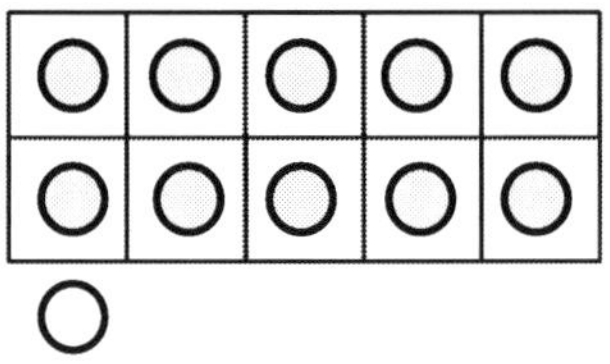

4) 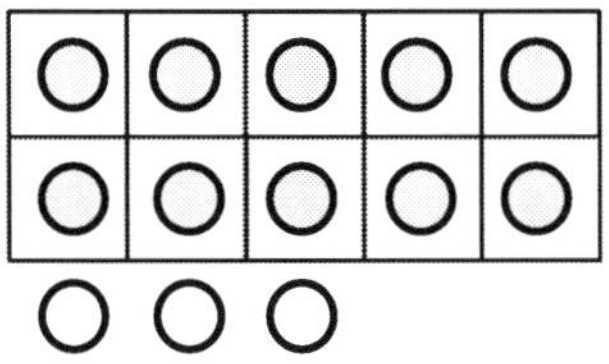

5) 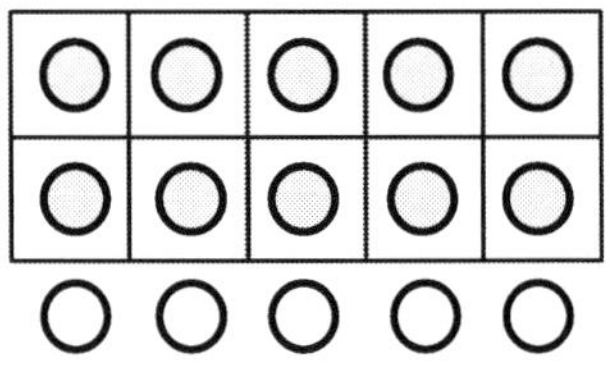

14

11

12

15

13

K.CC

Set 2: Match the counters to the expression.

1) 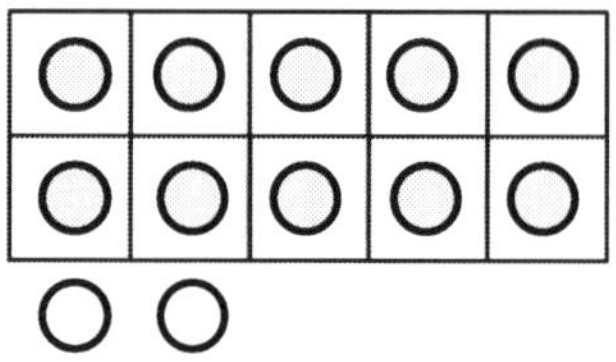

2) 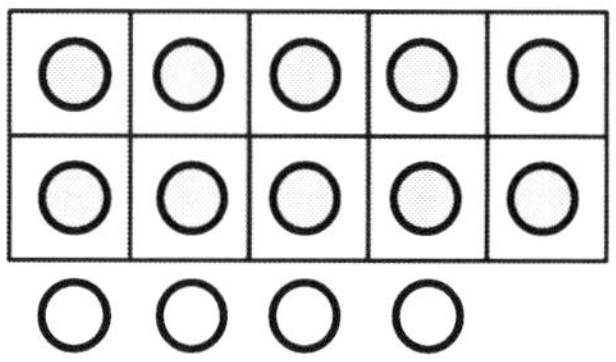

3) 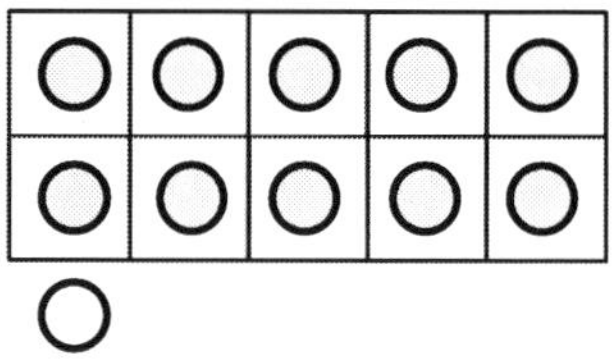

4) 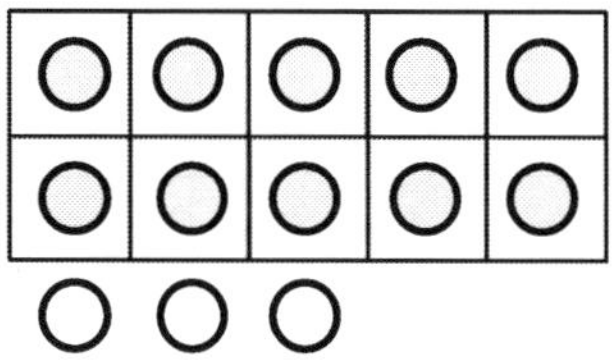

5) 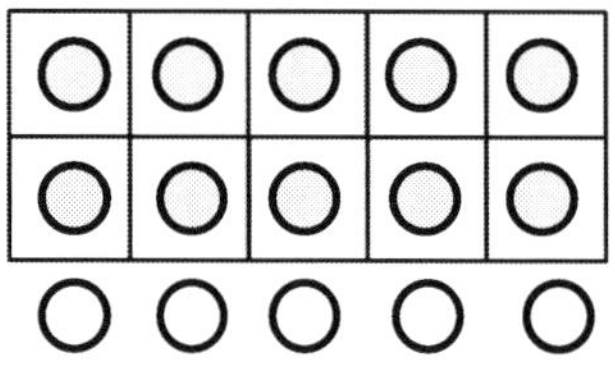

10 + 5

10 + 4

10 + 3

10 + 2

10 + 1

K.CC

Do you understand? **?** ✓

Unit Reflection

Numbers 0 to 20

Use this space to reflect on your understanding of the unit skills and concepts.

Skill/Concept	I can...	I need to work on...

Unit 7

Solid Shapes All Around Us

In this unit we will explore everyday objects that look like three-dimensional shapes. We will use attributes of flat and solid shapes to describe shapes. We will also compare weight and capacity.

Solid Shapes

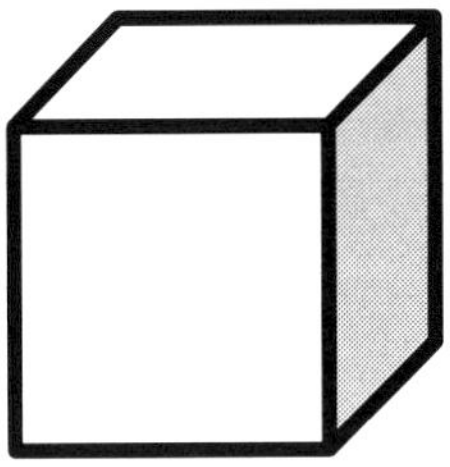

Cube

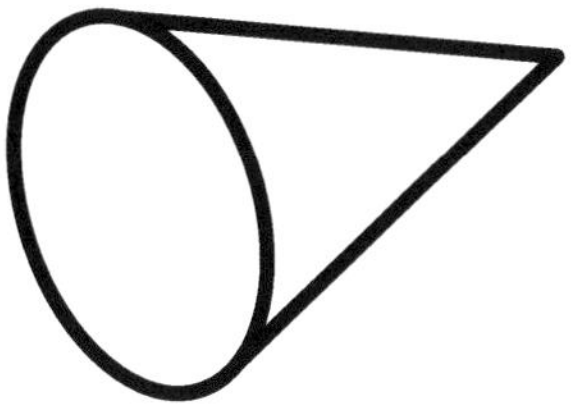

Cone

Cylinder

Sphere

Flat Shapes

Rectangle

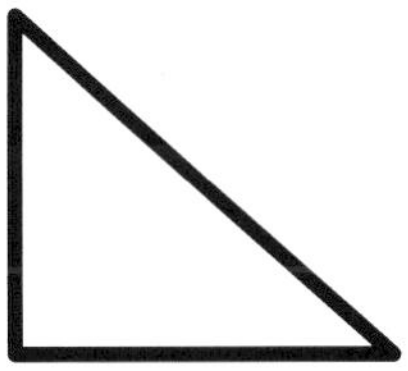

Triangle

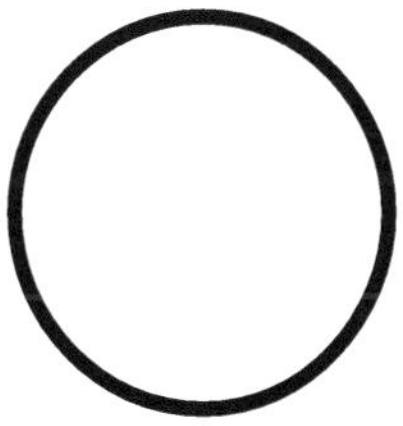

Circle

Square

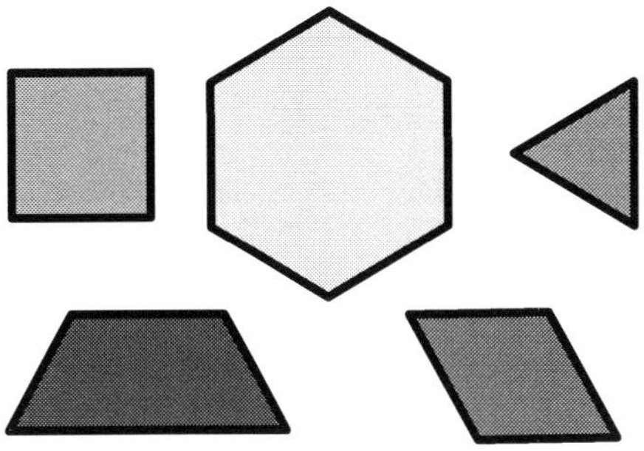

Pattern Blocks are Flat Shapes

A Ball Looks Like a Sphere

Unit Vocabulary

Solid Shapes All Around Us

Use this space to visualize the math vocabulary for this unit.

Word or Phrase	Visual Reminder

Unit Tools

Solid Shapes All Around Us

Use this space to visualize the tools for this unit.

Tools	This is a...	It is used to...

Affirmation: We are all smart in different ways.

Name: ______________________________ Date: ______________

Unit 7 Practice 1: I can identify missing shapes.

1) **Sample:** Circle the shape that is needed to fill the space.

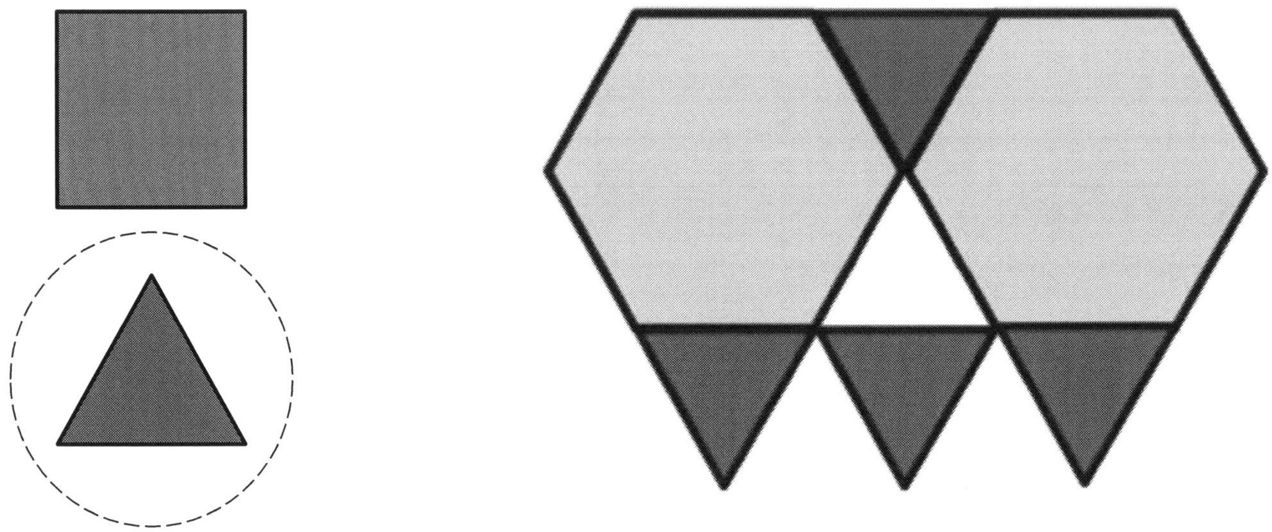

K.G.2

2) Circle the shape that is needed to fill the space.

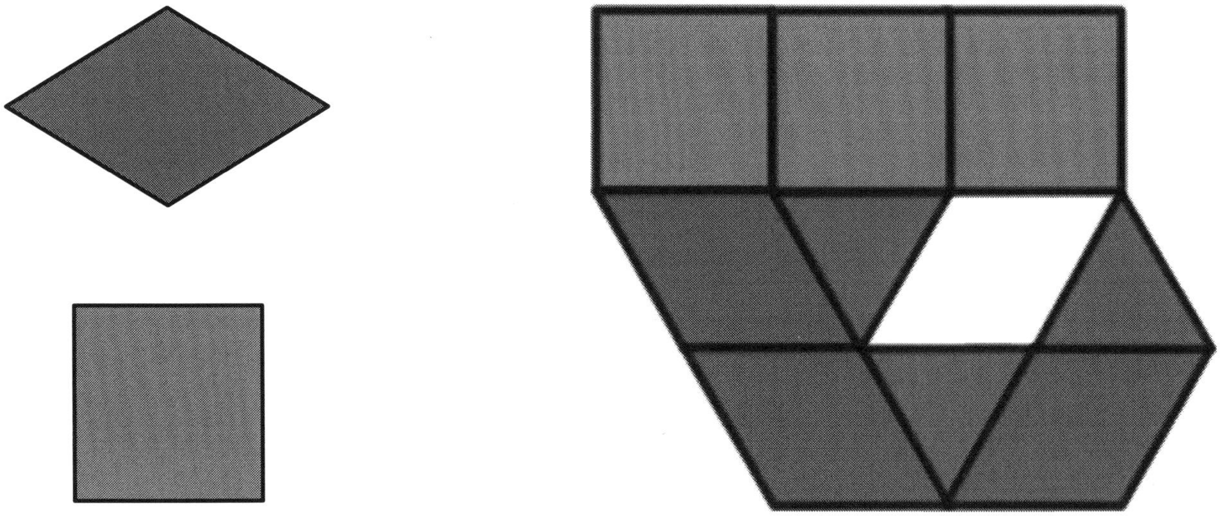

K.G.2

Do you understand? ? ✓

Affirmation: We are all smart in different ways.

Name: ______________________________ Date: ______________

Unit 7 Practice 1: I can identify missing shapes.

3) Circle the shape that is needed to fill the space.

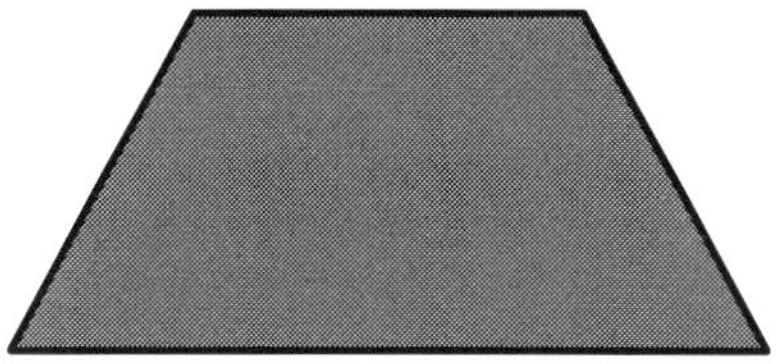

K.G.2

4) Circle the shape that is needed to fill the space.

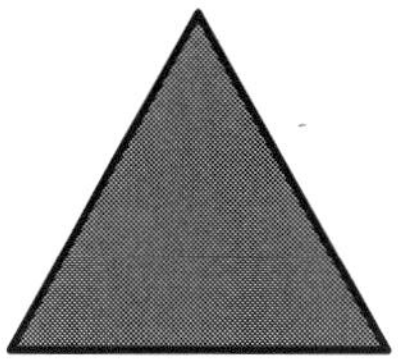

K.CC.5

Do you understand? ? ✓

Name: ______________________________ Date: ______________

Unit 7 Practice 2: I can identify more or less.

1) **Sample:** Circle the shape that is filled with **more** pattern blocks.

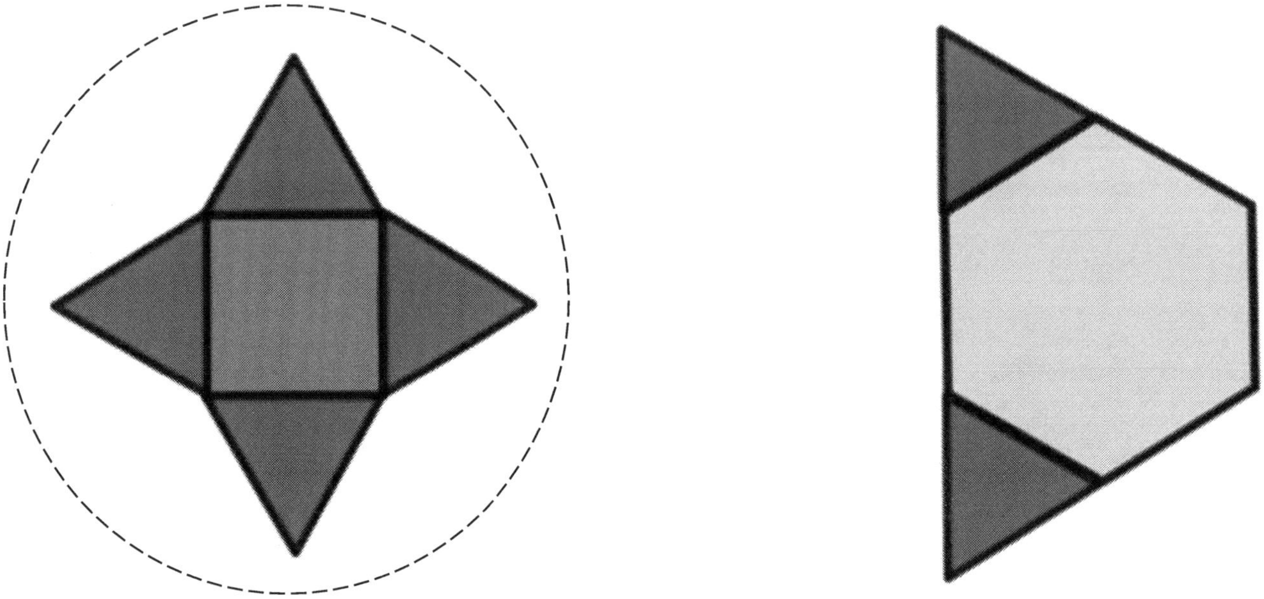

K.CC.6

2) Circle the shape that is filled with **more** pattern blocks.

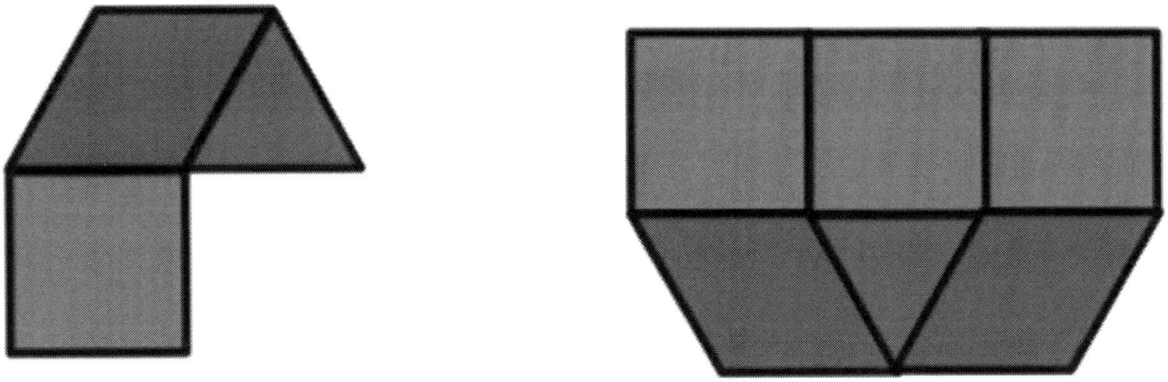

K.CC.6

Do you understand? **?** ✓

Name: ______________________________ Date: ______________

Unit 7 Practice 2: I can identify more or less.

3) Circle the shape that is filled with **fewer** pattern blocks.

K.CC.6

4) Circle the shape that is filled with **fewer** pattern blocks.

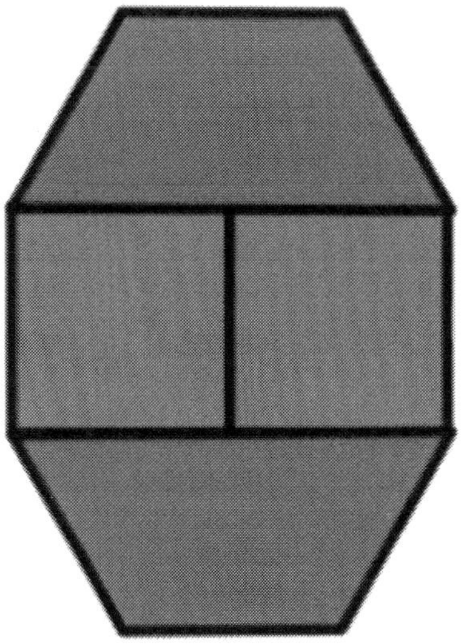

K.CC.6

Do you understand? ? ✓

Name: ______________________ Date: ____________

Unit 7 Practice 3: I can identify shapes.

1) **Sample:** How many △ triangles?

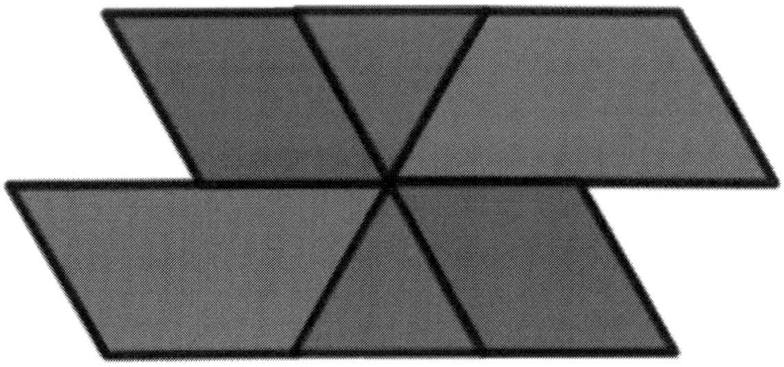

There are 2 triangles.

K.CC.5

2) How many □ squares?

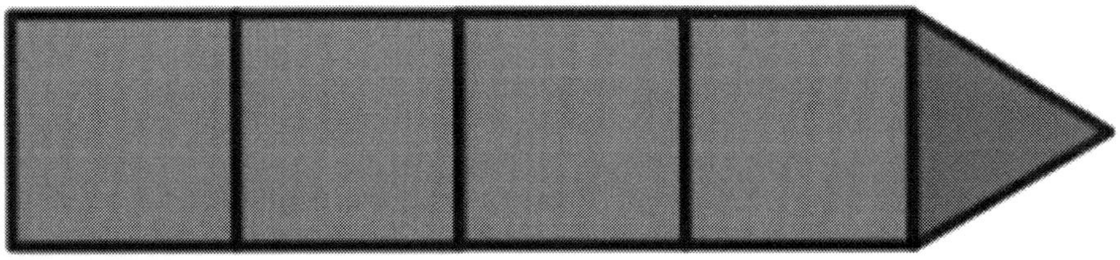

There are ________ squares.

K.CC.5

Do you understand? ? ✓

Affirmation: We are all smart in different ways.

Name: ______________________ Date: ____________

Unit 7 Practice 3: I can identify shapes.

3) How many △ triangles?

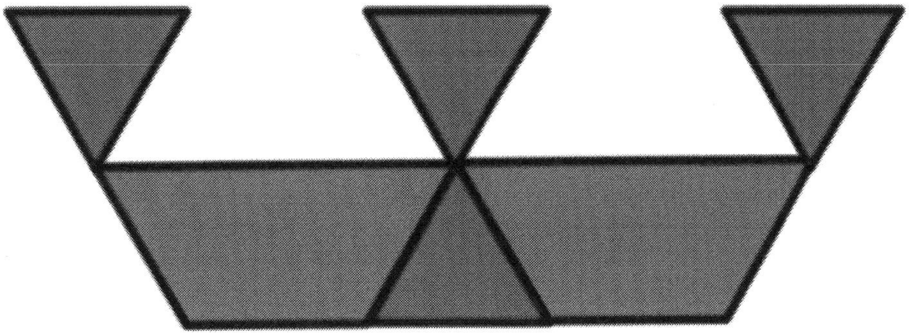

There are ________ triangles.

K.CC.5

4) How many ☐ squares?

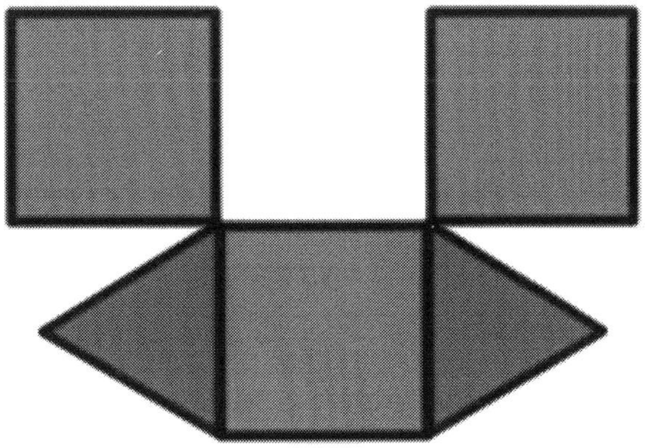

There are ________ squares.

K.CC.5

Do you understand? **?** ✓

Affirmation: We are all smart in different ways.

Name: ______________________________ Date: ______________

Unit 7 Practice 4: I can match shapes to expressions.

1) **Sample:** Juan used **2** pattern blocks.

Expression: 1 + 1

K.OA.1

2) Lisa used **4** pattern blocks.

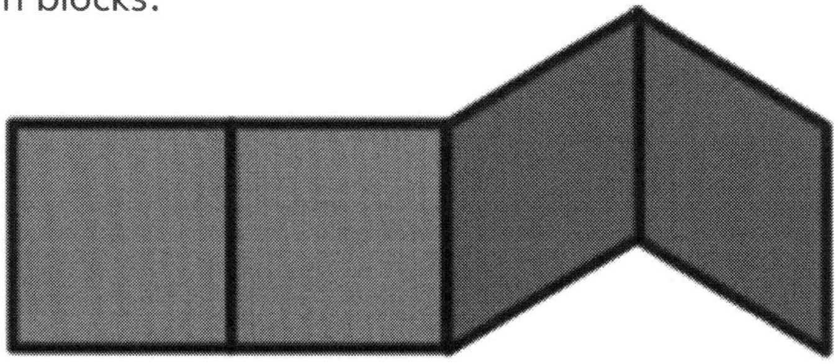

Expression: ______ + ______

K.OA.1

Do you understand? **?** ✓

Name: ______________________ Date: ____________

Unit 7 Practice 4: I can match shapes to expressions.

3) Diego used **5** pattern blocks.

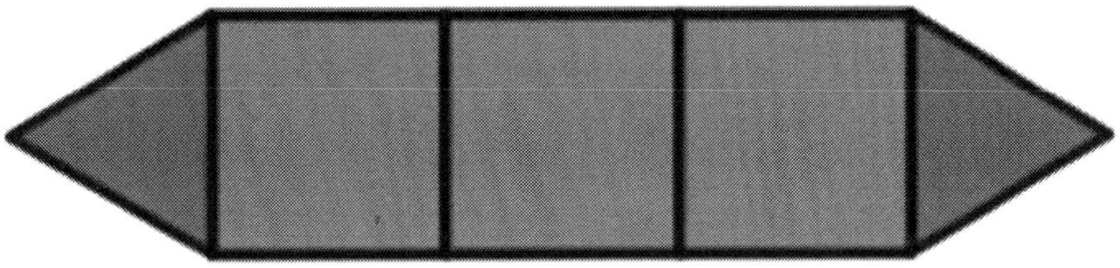

Expression: ______ + ______

K.OA.1

4) Anna used **7** pattern blocks.

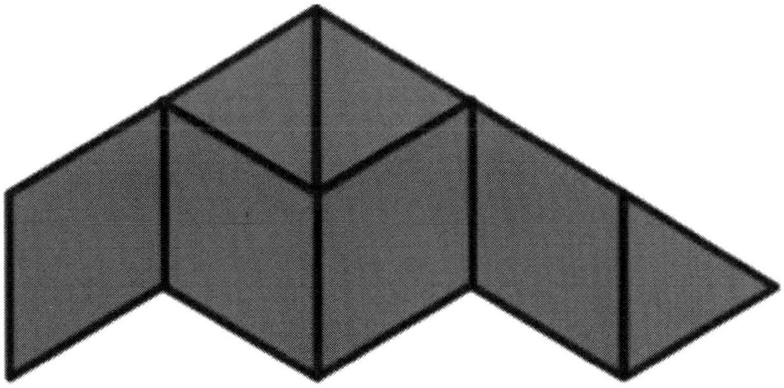

Expression: ______ + ______

K.OA.1

Do you understand? ? ✓

Name: ______________________ Date: ____________

Unit 7 Practice 5: I can match story problems to equations.

1) **Sample:** Mateo had **3** pattern blocks.
Brian gave him **4** more pattern blocks.
How many pattern blocks does Mateo have now?
Circle the equation that matches the story.

$3 + 4 = 7$ (circled) $7 - 4 = 3$

K.OA.1

2) Melissa had **2** pattern blocks.
Briana gave her **3** more pattern blocks.
How many pattern blocks does Melissa have now?
Circle the equation that matches the story.

$5 - 2 = 3$ $2 + 3 = 5$

K.OA.1

Do you understand? **?** ✓

Name: ______________________________ Date: ______________

Unit 7 Practice 5: I can match story problems to equations.

3) Nina had **6** pattern blocks.
Kiara gave her **3** more pattern blocks.
How many pattern blocks does Nina have now?
Circle the equation that matches the story.

$6 + 3 = 9$ $9 - 6 = 3$

K.OA.1

4) Theo had **4** pattern blocks.
Joseph gave him **3** more pattern blocks.
How many pattern blocks does Theo have now?
Circle the equation that matches the story.

$7 - 4 = 3$ $4 + 3 = 7$

K.OA.1

Do you understand? **?** ✓

Name: ________________________________ Date: ______________

Unit 7 Practice 6: I can make 10.

1) **Sample:** Jack built a shape with **10** pattern blocks.
He only used rhombuses and trapezoids.
How many of each pattern block did he use.

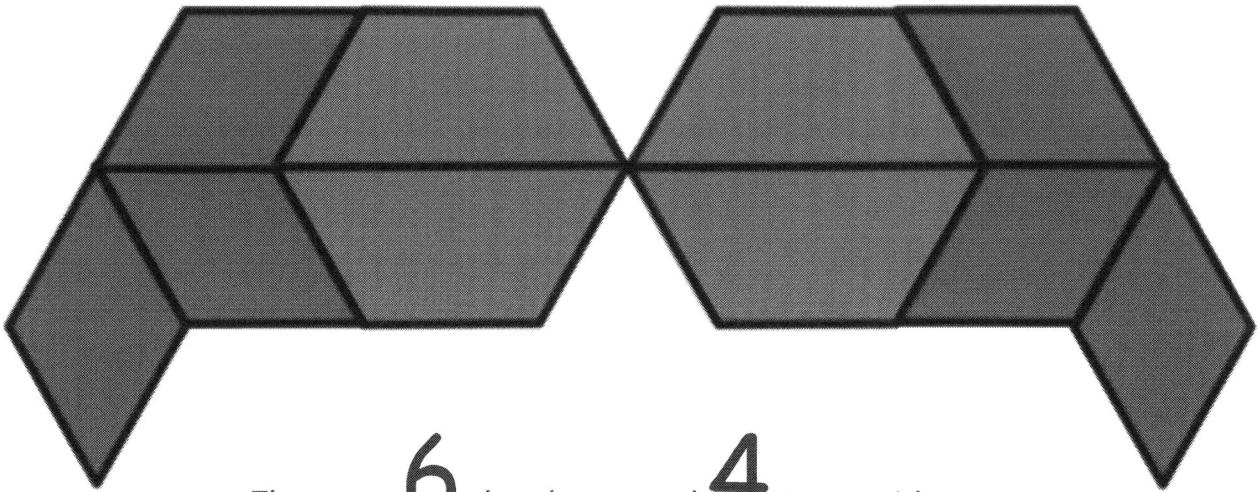

There are 6 rhombuses and 4 trapezoids.

K.OA.3

2) Oliver built a shape with **10** pattern blocks.
He only used rhombuses and triangles.
How many of each pattern block did he use?

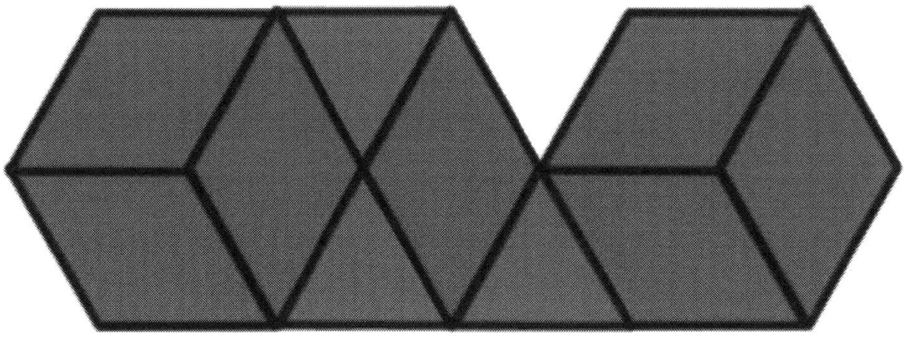

There are ________ rhombuses and ________ triangles.

K.OA.3

Do you understand? **?** ✓

Name: ______________________ Date: __________

Unit 7 Practice 6: I can make 10.

3) Isabella built a shape with **10** pattern blocks.
She only used rhombuses and trapezoids.
How many of each pattern block did he use?

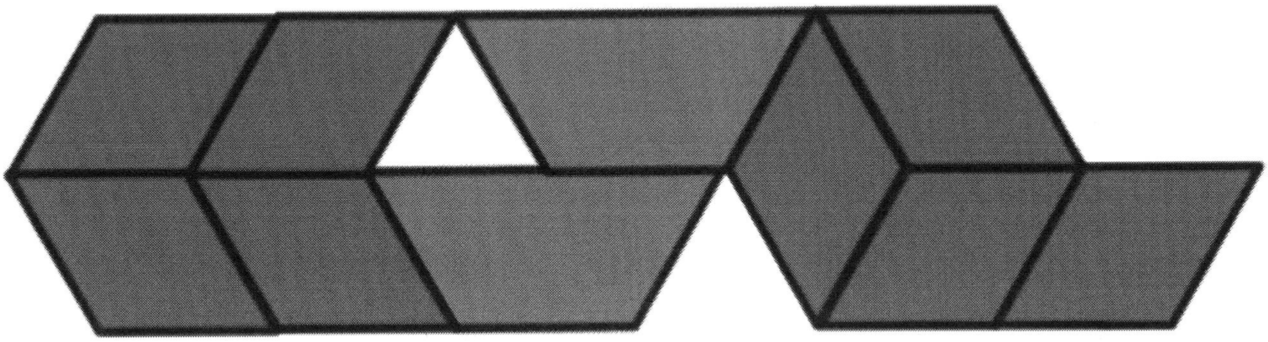

There are ________ rhombuses and ________ trapezoids.

K.OA.3

4) Emily built a shape with **10** pattern blocks.
She only used rhombuses and triangles.
How many of each pattern block did he use?

There are ________ rhombus and ________ triangles.

K.OA.3

Do you understand? **?** ✓

Name: ______________________ Date: __________

Unit 7 Practice 7: I can identify shapes.

1) **Sample:** Put a check mark ✓ on the shape that is **flat, NOT solid**.

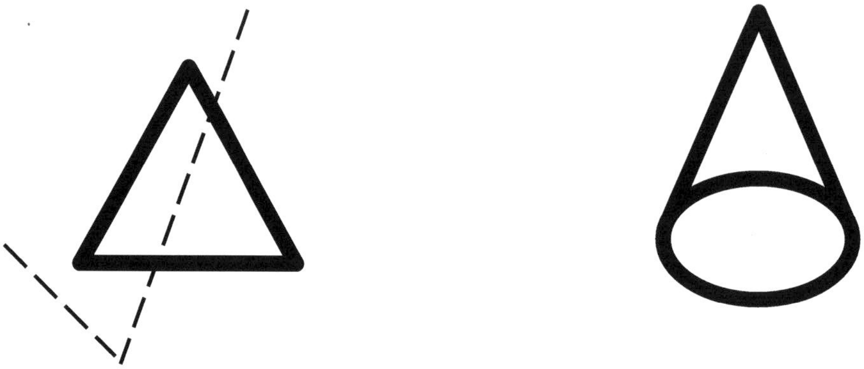

K.G.4

2) Put a check mark ✓ on the shape that is **flat, NOT solid**.

K.G.4

Do you understand? **?** ✓

Name: ______________________ Date: ____________

Unit 7 Practice 7: I can identify shapes.

3) Put a check mark ✓ on the shape that is **flat, NOT solid**.

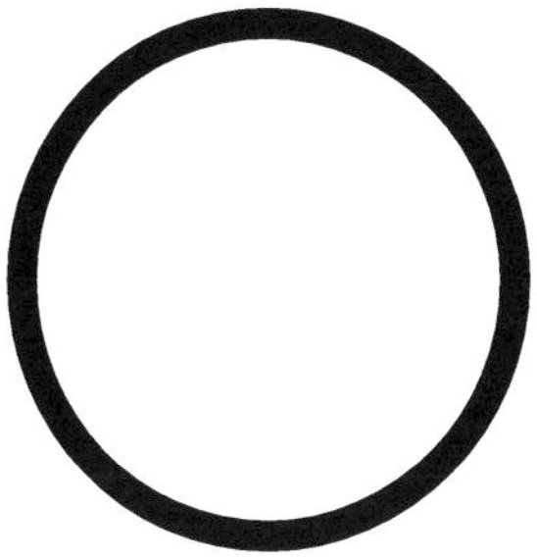

K.G.4

4) Put a check mark ✓ on the shape that is **flat, NOT solid**.

K.G.4

Do you understand? **?** ✓

Name: ______________________________ Date: ______________

Unit 7 Practice 8: I can tell heavier or lighter.

1) **Sample:** Put a check mark ✓ on the object that is **heavier**.
Put an X on the object that is **lighter**.

K.MD.2

2) Put a check mark ✓ on the object that is **heavier**.
Put an X on the object that is **lighter**.

K.MD.2

Do you understand? **?** ✓

Name: ______________________________ Date: ______________

Unit 7 Practice 8: I can tell heavier or lighter.

3) Put a check mark ✓ on the object that is **heavier**.
Put an X on the object that is **lighter**.

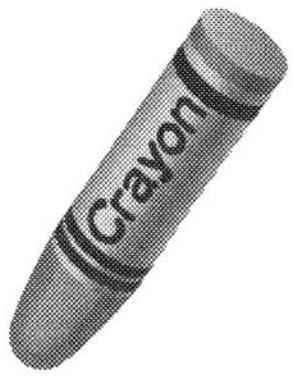

K.MD.2

4) Put a check mark ✓ on the object that is **heavier**.
Put an X on the object that is **lighter**.

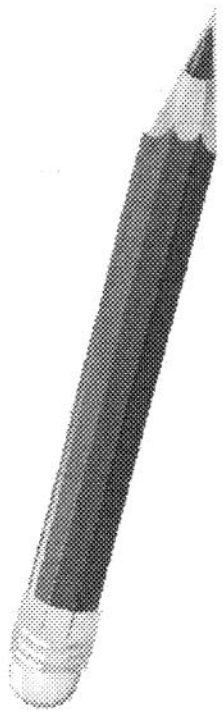

K.MD.2

Do you understand? **?** ✓

Name: ______________________________ Date: ______________

Unit 7 Practice 9: I can tell which holds more or less.

1) **Sample:** Put a check mark ✓ on the object that holds **more**.
Put an X on the object that holds **less**.

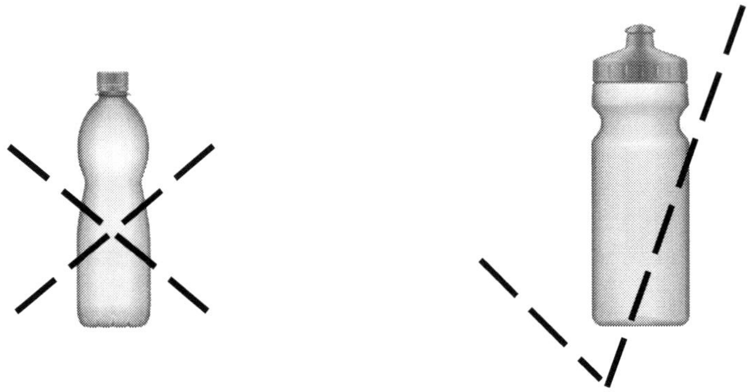

K.MD.2

2) Put a check mark ✓ on the object that holds **more**.
Put an X on the object that holds **less**.

K.MD.2

Do you understand? **?** ✓

Name: ______________________________ Date: ______________

Unit 7 Practice 9: I can tell which holds more or less.

3) Put a check mark ✓ on the object that holds **more**.
Put an X on the object that holds **less**.

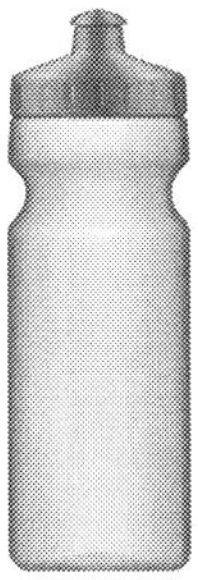

K.MD.2

4) Put a check mark ✓ on the object that holds **more**.
Put an X on the object that holds **less**.

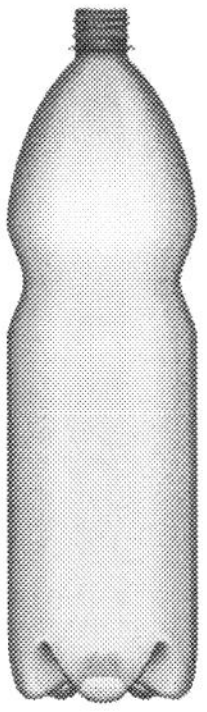

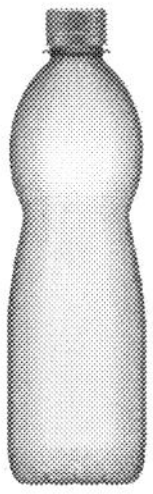

K.MD.2

Do you understand? **?** ✓

Affirmation: Making mistakes is how I learn new things.

Name: ______________________________ Date: ______________

Unit 7 Practice 10: I can identify solid shapes.

1) **Sample:** Put a check mark ✓ on the object that looks like a **cone**.

K.G.4

2) Put a check mark ✓ on the object that looks like a **cylinder**.

K.G.4

Do you understand? **?** ✓

Name: ______________________________ Date: ______________

Unit 7 Practice 10: I can identify solid shapes.

3) Put a check mark ✓ on the object that looks like a cube.

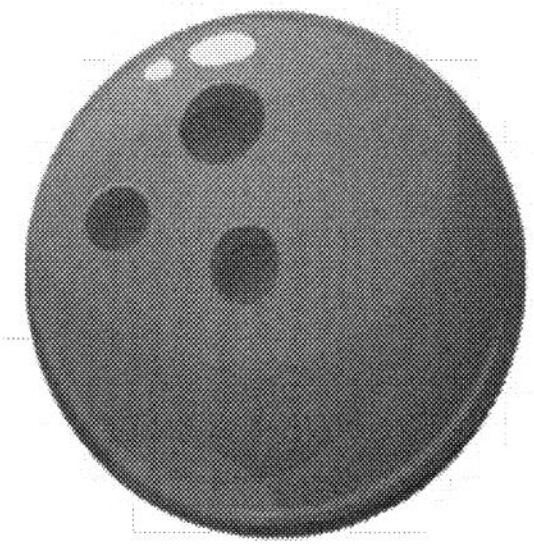

K.G.4

4) Put a check mark ✓ on the object that looks like a **sphere**.

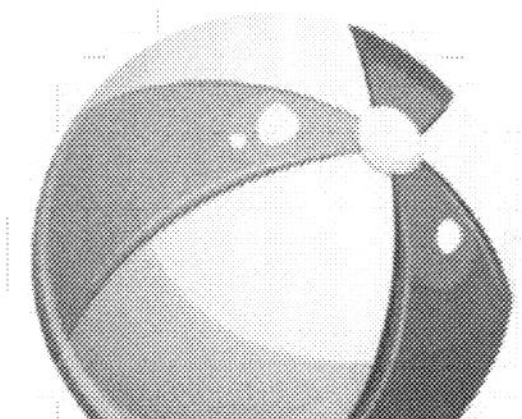

K.G.4

Do you understand? **?** ✓

Name: ______________________ Date: ____________

Unit 7 Practice 11: I can describe shapes.

1) **Sample:** Put a check mark ✓ on the shape that has **corners**.
Put an X on the shape that has **NO corners**.

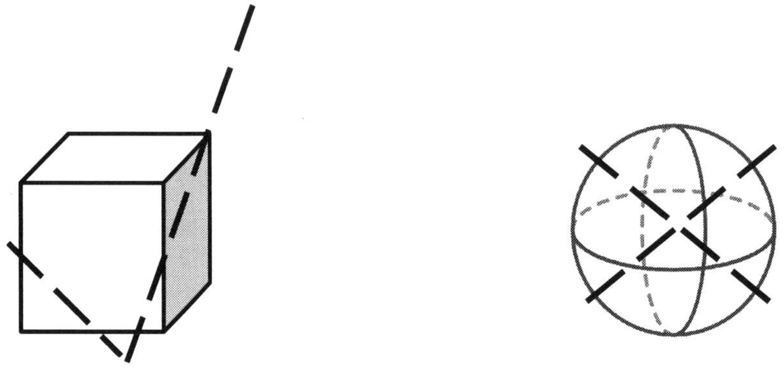

K.G.4

2) Put a check mark ✓ on the shape that can **roll**.
Put an X on the shape that has **CANNOT roll**.

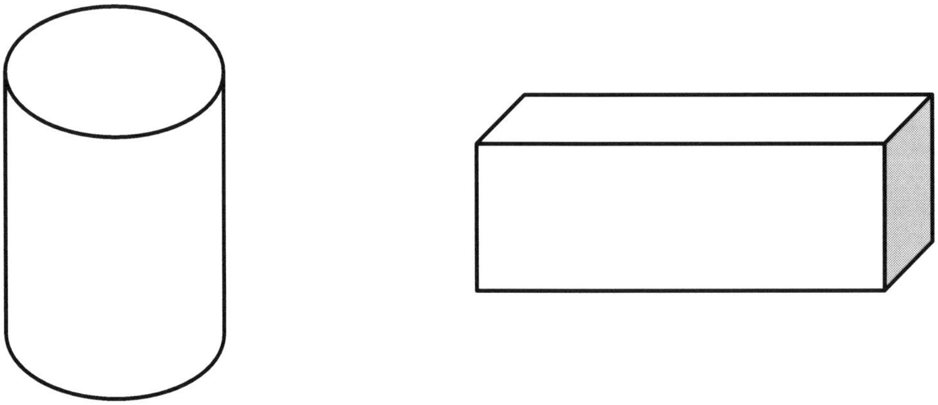

K.G.4

Do you understand? **?** ✓

Name: ______________________________ Date: ______________

Unit 7 Practice 11: I can describe shapes.

3) Put a check mark ✓ on the shape that has **no corners**.
Put an X on the shape that has **NO corners**.

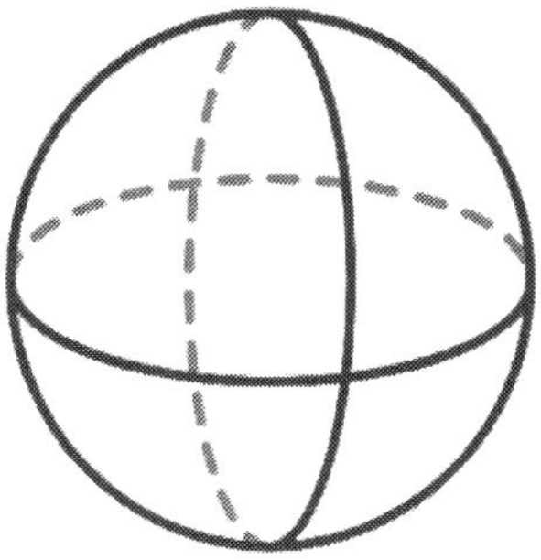

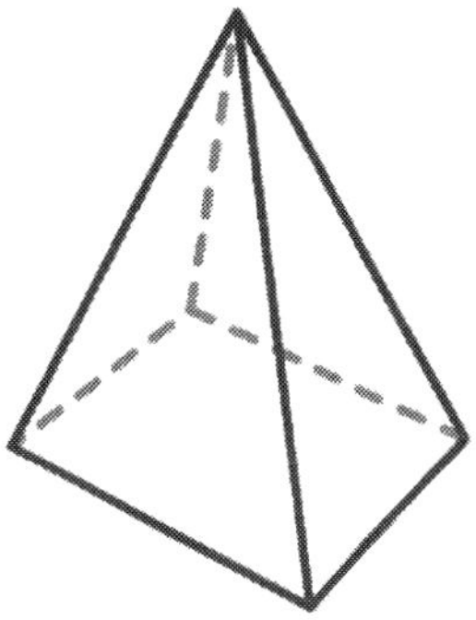

K.G.4

4) Put a check mark ✓ on the shape that can **roll**.
Put an X on the shape that has **CANNOT roll**.

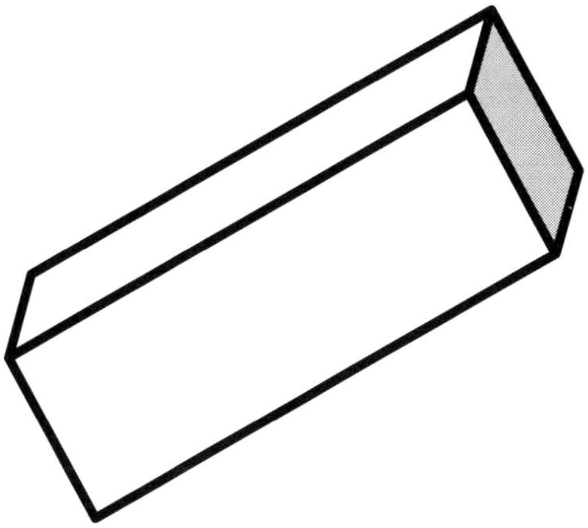

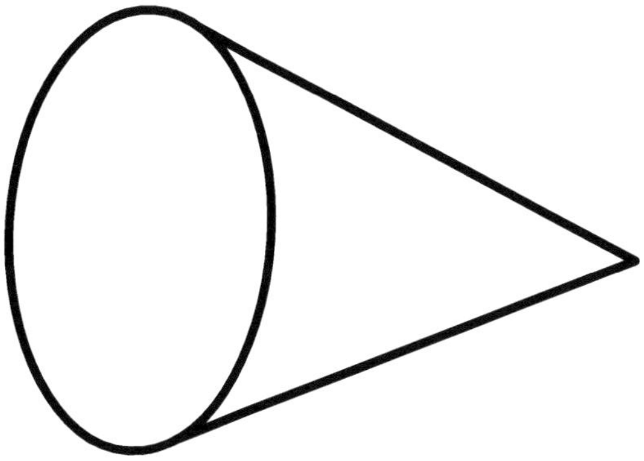

K.G.4

Do you understand? **?** ✓

Affirmation: I am brilliant, bright, and getting better every day!

Name: ______________________ Date: ______________

Unit 7 Practice 12: I can identify solid shapes.

1) **Sample:** Put a check mark ✓ on the **solid** shape.
Put an X on the **flat** shape.

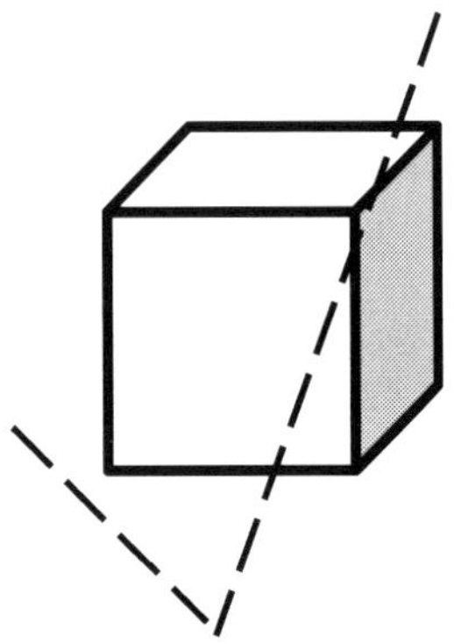

K.G.3

2) Put a check mark ✓ on the **solid** shape.
Put an X on the **flat** shape.

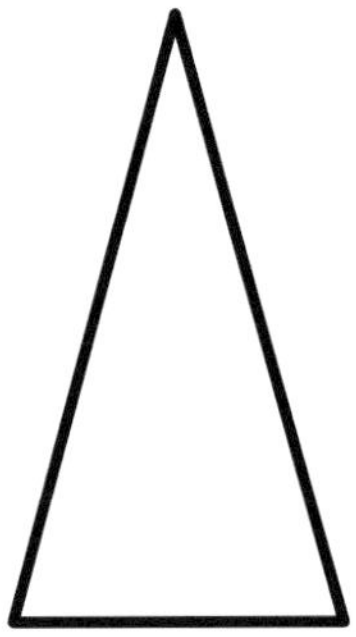

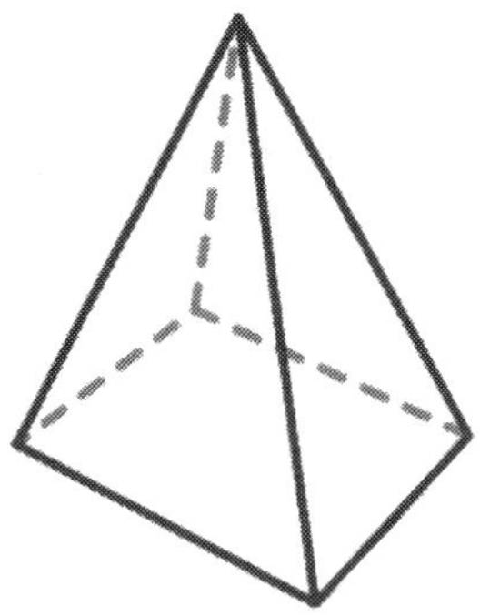

K.G.3

Do you understand? **?** ✓

Name: ______________________________ Date: ______________

Unit 7 Practice 12: I can identify solid shapes.

3) Put a check mark ✓ on the **solid** shape.

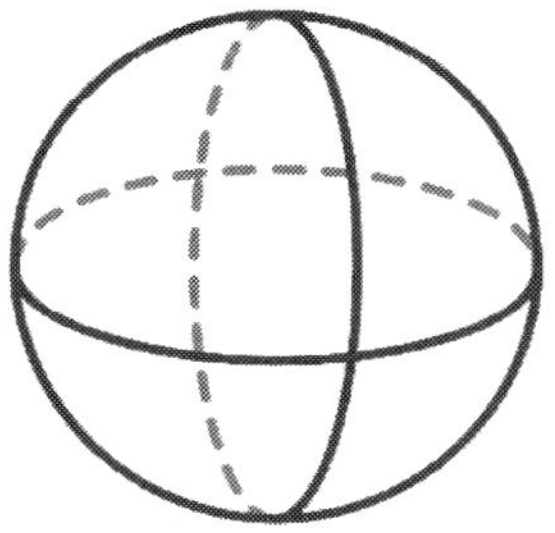

K.G.3

4) Put a check mark ✓ on the **solid** shape.

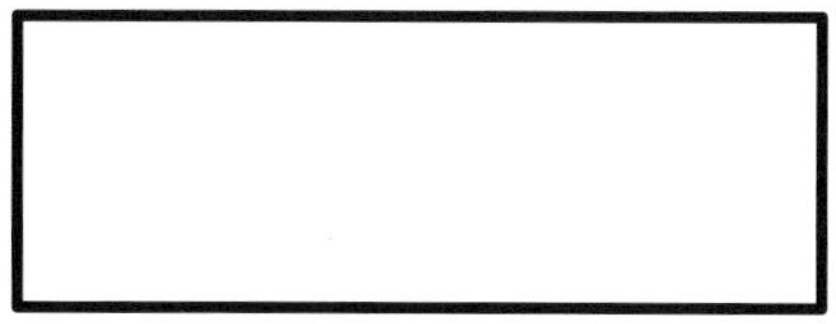

K.G.3

Do you understand? **?** ✓

Affirmation: I am brilliant, bright, and getting better every day!

Name: ______________________________ Date: ______________

Unit 7 Practice 13: I can identify solid shapes.

1) Put an X on the objects that looks like a **cube**.

I see 4 cubes.

K.G.1

2 Put an X on the the objects that looks like a **sphere**.

I see ______ spheres.

K.G.1

Do you understand? ? ✓

Name: ______________________ Date: __________

Unit 7 Practice 13: I can identify solid shapes.

3) Put an X on the objects that looks like a **cone**.

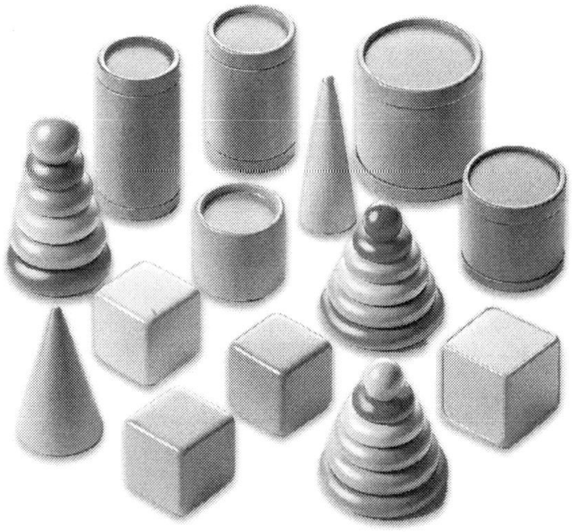

I see ________ cones.

K.G.1

4) Put an X on the objects that looks like a **cylinder**.

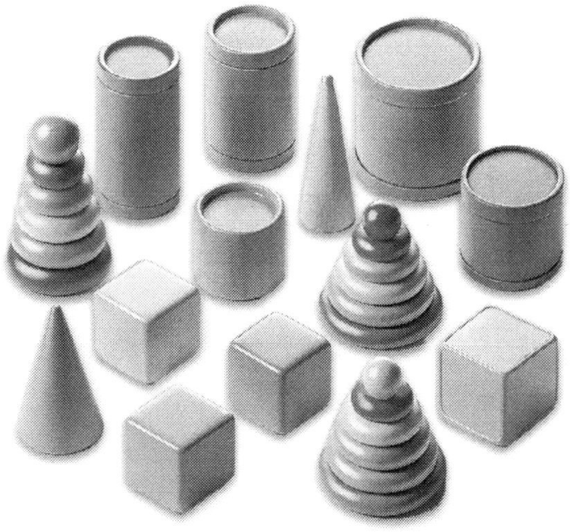

I see ________ cylinders.

K.G.1

Do you understand? **?** ✓

Affirmation: I am brilliant, bright, and getting better every day!

Name: ______________________________ Date: ______________

Unit 7 Practice 14: I can talk about shapes.

1) **Sample:** Where is the  cylinder?

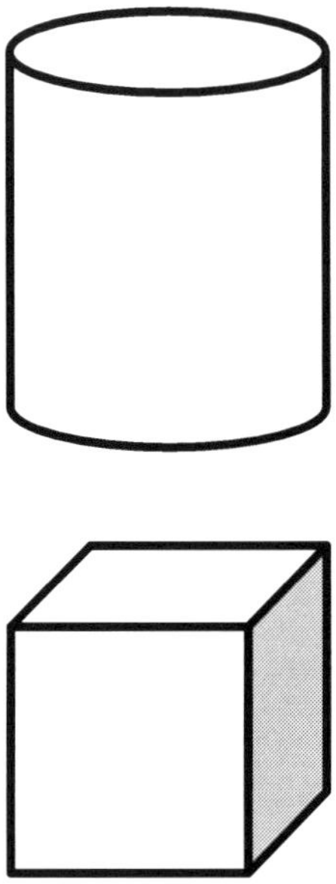

above

The cylinder is ____below____ the cube.

K.G.1

2) Where is the  cube?

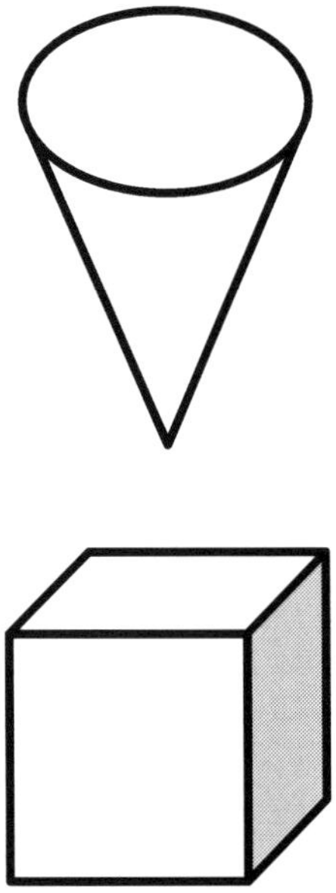

above

The cube is ____below____ the cone.

K.G.1

Do you understand? ? ✓

Name: ______________________________ Date: ______________

Unit 7 Practice 14: I can talk about shapes.

3 Where is the  cylinder?

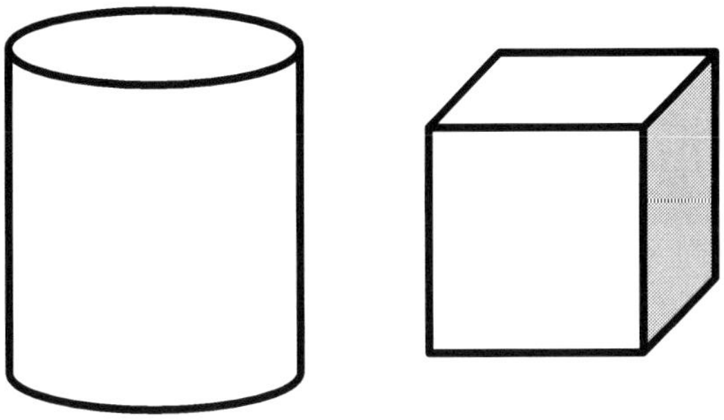

above

The cylinder is **next to** the cube.

K.G.1

4) Where is the  cube?

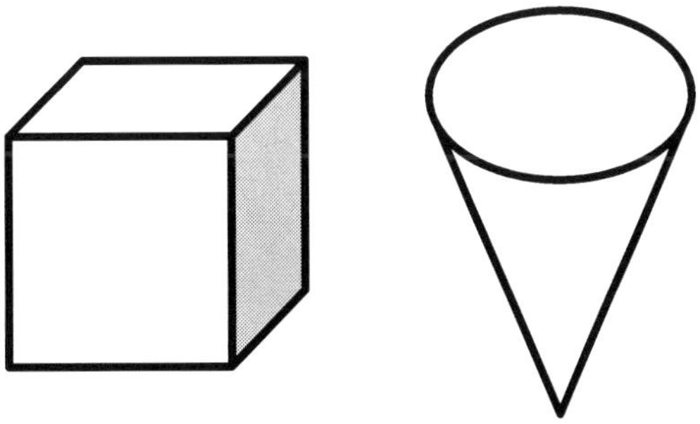

next to

The cube is **below** the cone.

K.G.1

Do you understand? **?** ✓

Name: ______________________ Date: ____________

Unit 7 Practice 15: I can count solid shapes.

1) **Sample:** Count the solid shapes.

7 cones on top

3 cylinders on bottom

10 shapes in all

K.G.1

2) Count the solid shapes.

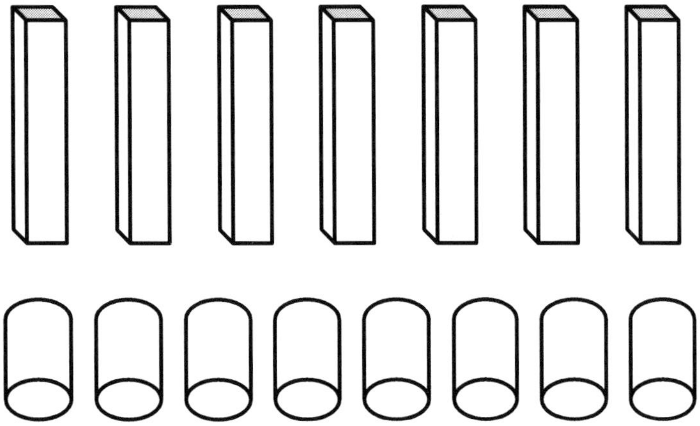

______ prisms on top

______ cylinders on bottom

______ shapes in all

K.G.1

Do you understand? ? ✓

Affirmation: I am brilliant, bright, and getting better every day!

Name: ______________________________ Date: ______________

Unit 7 Practice 15: I can count solid shapes.

3 Count the solid shapes.

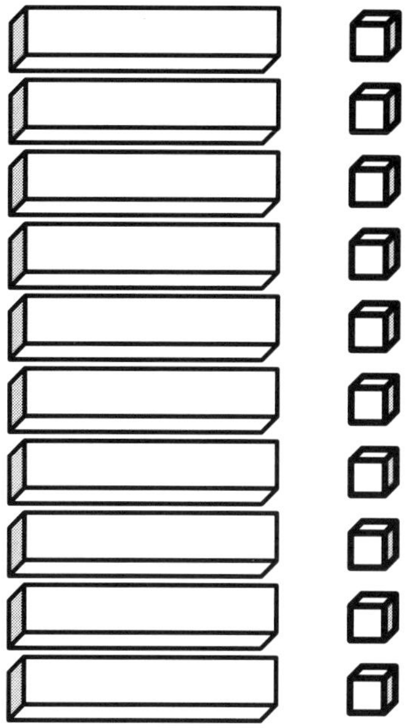

_______ prisms on the left

_______ cubes on the right

_______ shapes in all

K.G.1

4) Count the solid shapes.

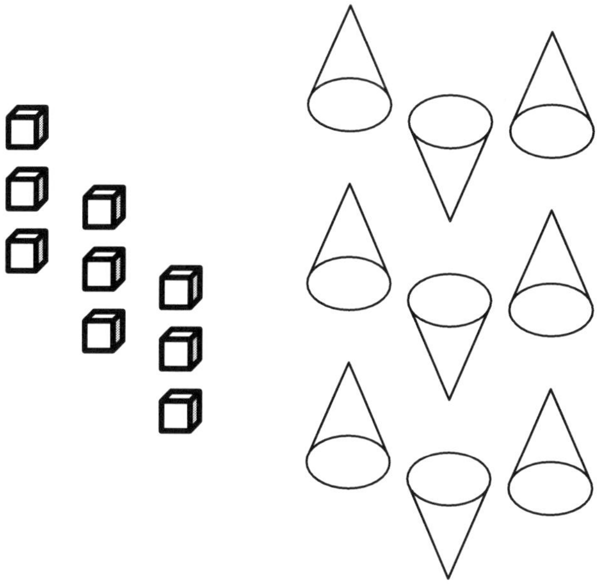

_______ cubes on the left

_______ cones on the right

_______ shapes in all

K.G.1

Do you understand? **?** ✓

Name: ______________________________ Date: ______________

Unit 7 Practice 16: I can match shapes to pictures.

1) **Sample:** Match the pictures to the shapes.

sphere

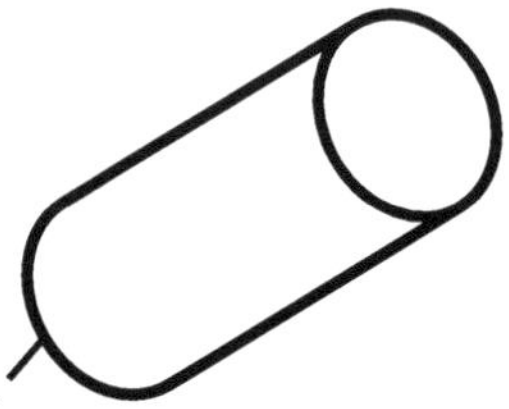

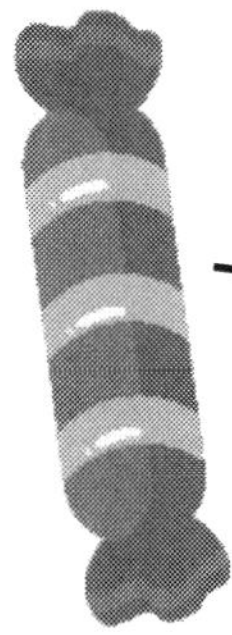

cylinder

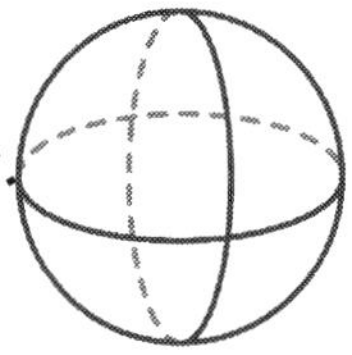

K.G.1

2) Match the pictures to the shapes.

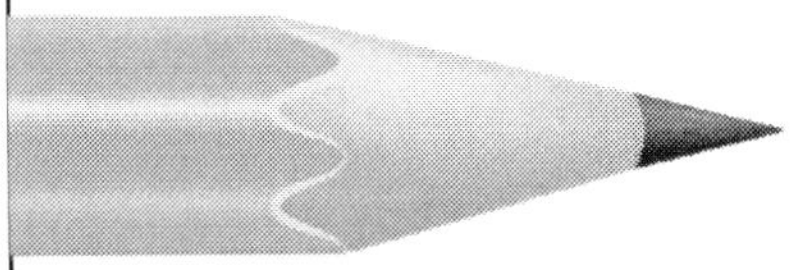

cube

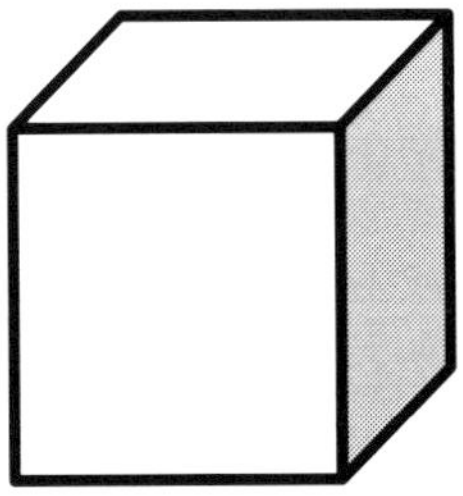

cone

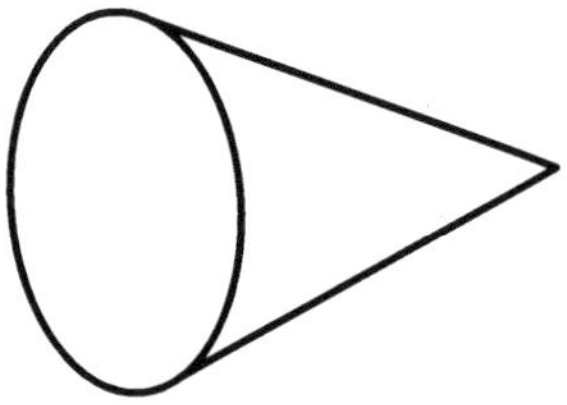

K.G.1

Do you understand? ? ✓

Name: ______________________________ Date: ______________

Unit 7 Practice 16: I can match shapes to pictures.

3) Match the pictures to the shapes.

prism

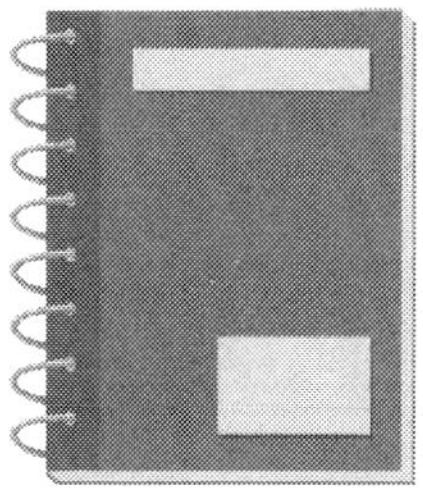

cone

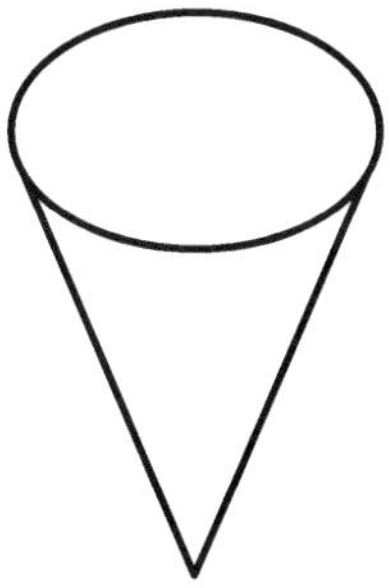

K.G.1

4) Match the pictures to the shapes.

cylinder

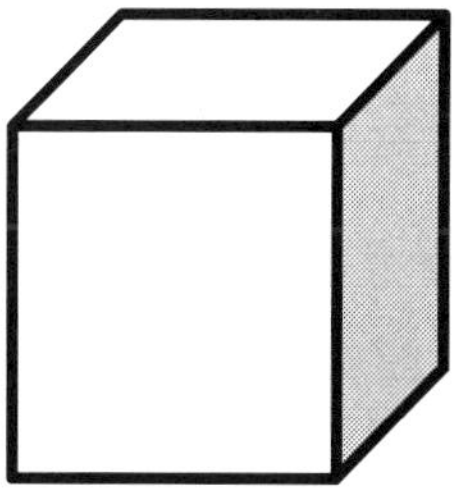

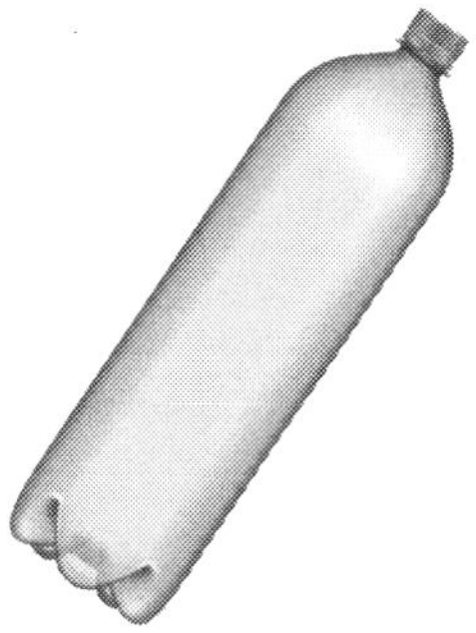

cube

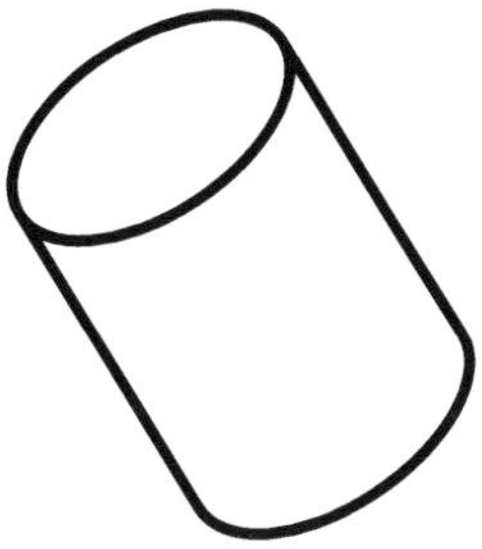

K.G.1

Do you understand? ? ✓

Name: ______________________ Date: ____________

I can practice grade level fluencies.

Set 1: Write the number of blocks.

1) 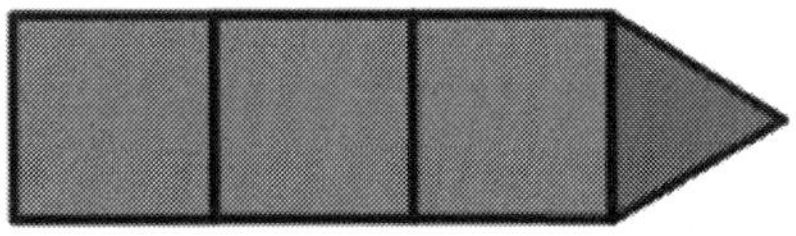 4

2) 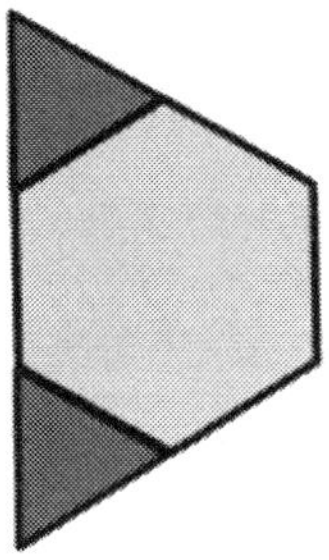

3) 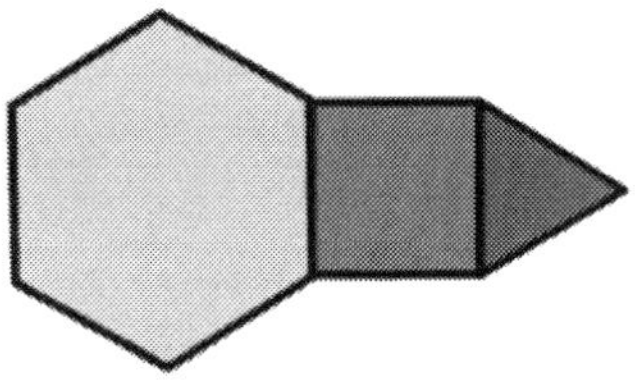

4) 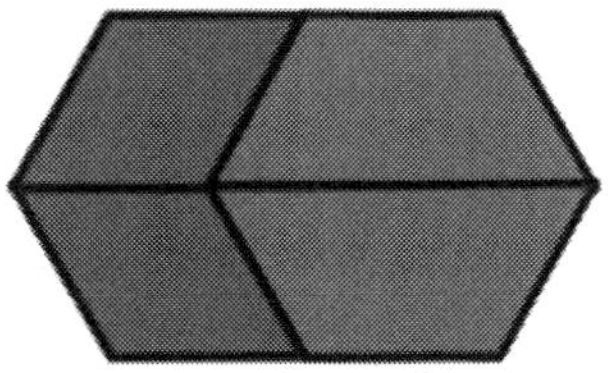

5) 

K.CC

Set 2: Match the shape to the picture.

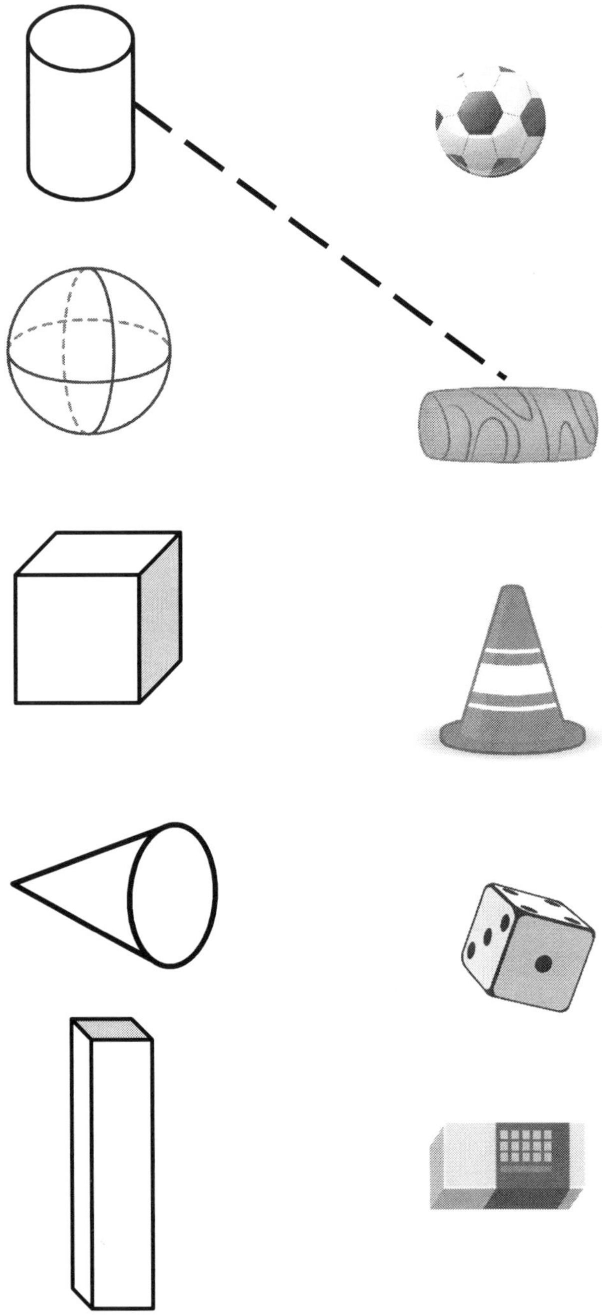

1)
2)
3)
4)
5)

K.CC

Do you understand? ? ✓

Unit Reflection

Solid Shapes All Around Us

Use this space to reflect on your understanding of the unit skills and concepts.

Skill/Concept	I can...	I need to work on...

Unit 8

Putting It All Together

In this unit, we will revisit working with addition and subtraction. We will compose and decompose numbers and solve story problems.

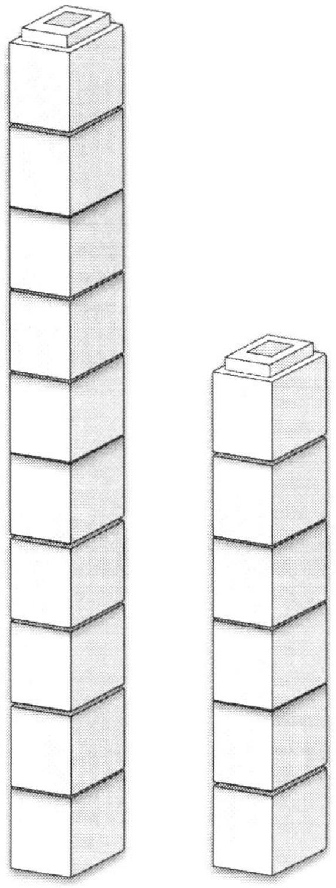

10 + 6 = 16

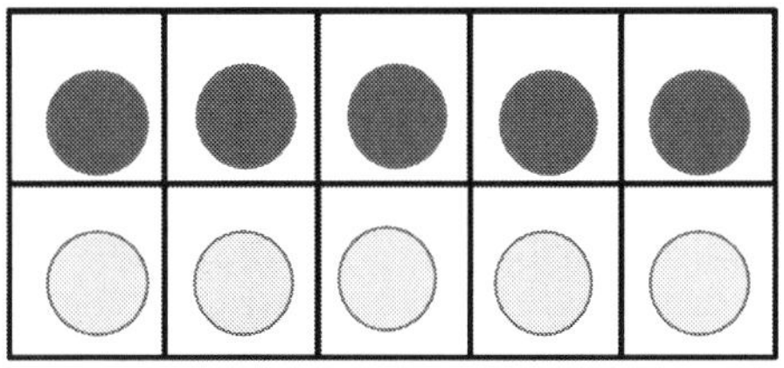

10 = 5 + 5

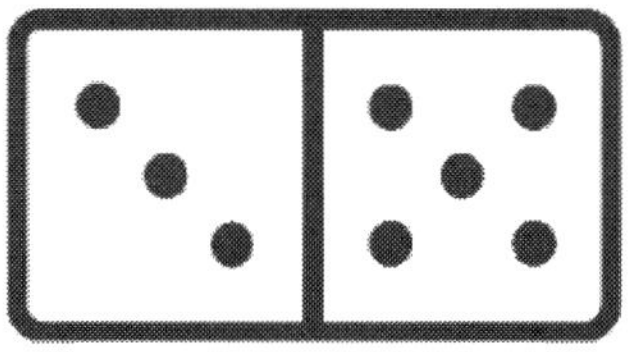

3 + 5

Story Problem:

Paul has 2 △ triangles.

Paul has 3 □ squares.

How many shapes does Paul have?

2 + 3 = 5

Paul has 5 shapes.

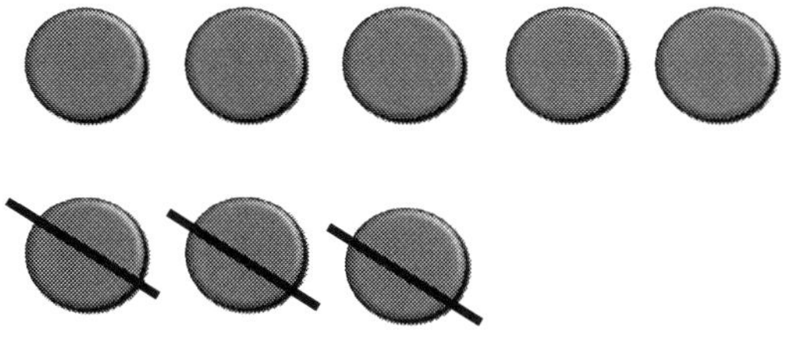

8 - 3

Unit Vocabulary

Putting It All Together

Use this space to visualize the math vocabulary for this unit.

Word or Phrase	Visual Reminder

Unit Tools

Putting It All Together

Use this space to visualize the tools for this unit.

Tools	This is a...	It is used to...

Affirmation: I don't give up!

Name: ______________________________ Date: ______________

Unit 8 Practice 1: I can tell more or less.

1) **Sample:** Put a check mark ✓ on the collection that has **more**.
Put an X on the collection that has **less**.

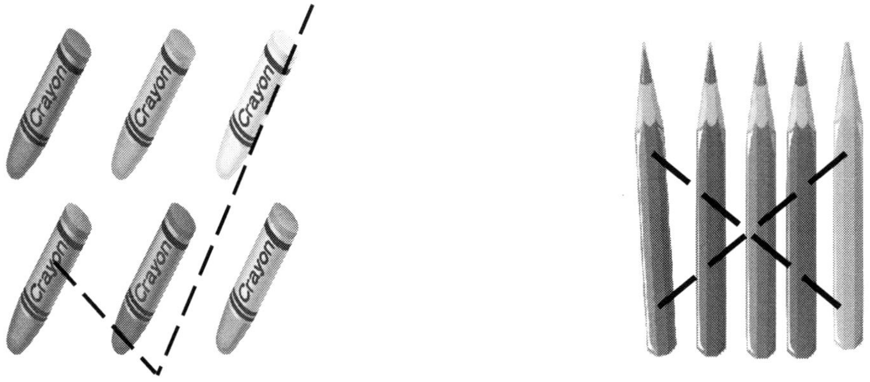

K.CC.6

2) Put a check mark ✓ on the collection that has **more**.
Put an X on the collection that has **less**.

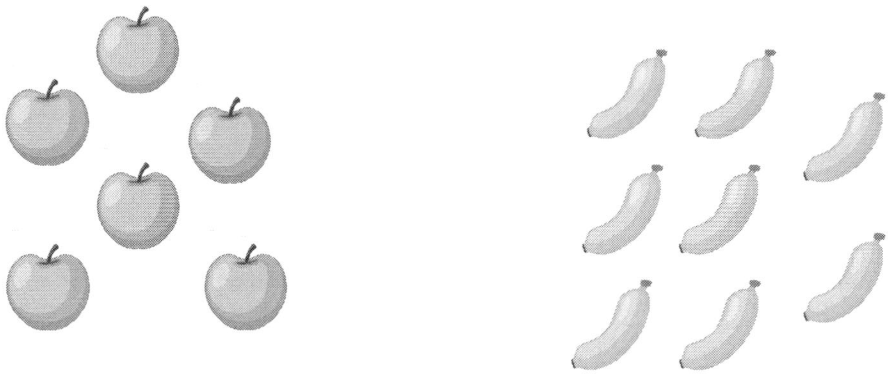

K.CC.6

Do you understand? **?** ✓

Name: ______________________________ Date: ______________

Unit 8 Practice 1: I can tell more or less.

3) Put a check mark ✓ on the collection that has **more**.
Put an X on the collection that has **less**.

K.CC.6

4) Put a check mark ✓ on the collection that has **less**.
Put an X on the collection that has **less**.

K.CC.6

Do you understand? **?** ✓

Name: ______________________________ Date: ____________

Unit 8 Practice 2: I can count objects.

1) **Sample:** How many circles?

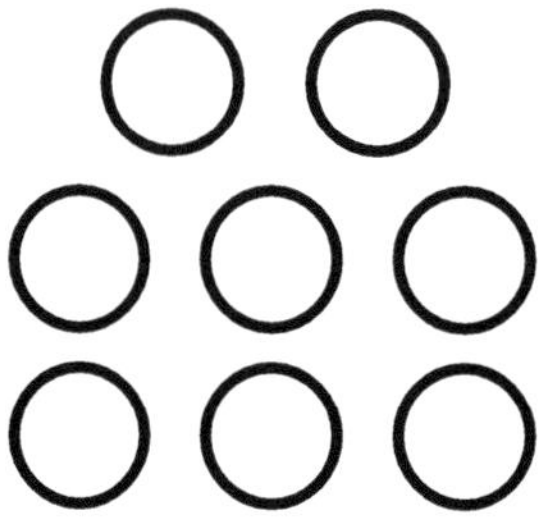

8 circles

K.CC.5

2) How many bears?

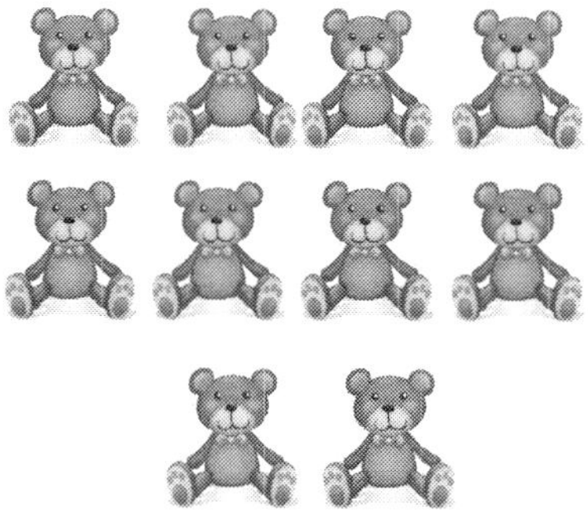

______ bears

K.CC.5

Do you understand? **?** ✓

Name: ______________________ Date: ____________

Unit 8 Practice 2: I can count objects.

3) How many crayons?

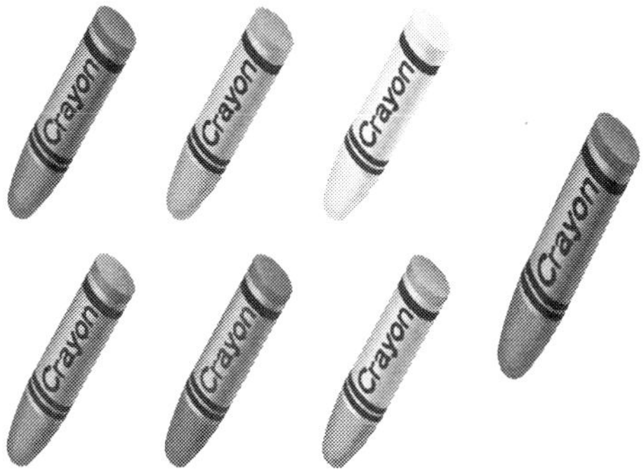

______ crayons

K.CC.5

4) How many paper clips?

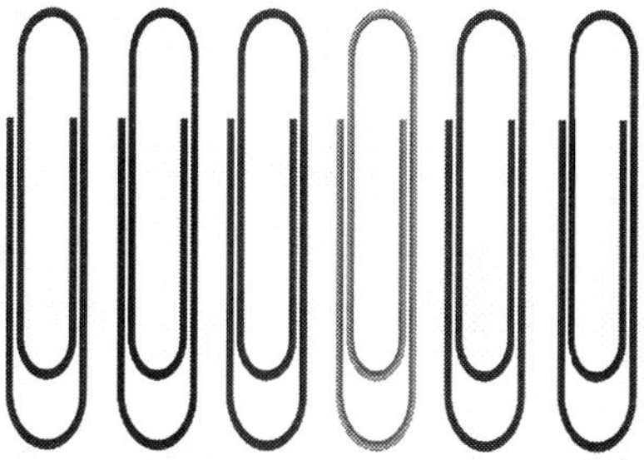

______ paper clips

K.CC.5

Do you understand? ? ✓

Name: ______________________ Date: ____________

Unit 8 Practice 3: I can add and subtract.

1) **Sample:** There were **6** flies.
2 flies joined them.
How many flies now?

6 + 2

There are 8 flies.

K.OA.2

2) There were **5** spiders.
2 spiders joined them.
How many spiders now?

_____ + _____

There are _______ spiders.

K.OA.2

Do you understand? ? ✓

Name: ____________________ Date: __________

Unit 8 Practice 3: I can add and subtract.

3) There were **6** butterflies.
2 butterflies flew away.
How many butterflies now?

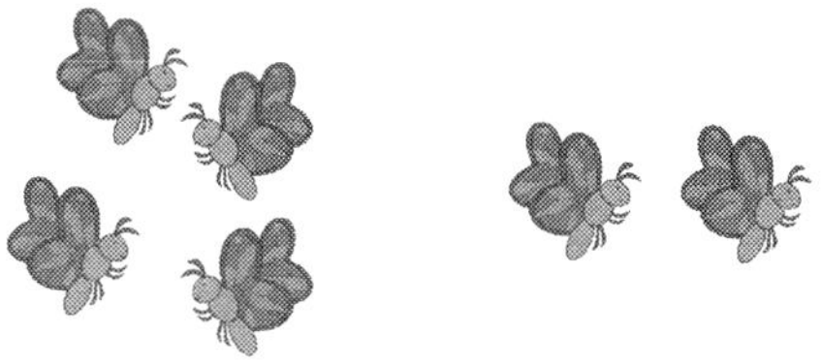

______ − ______

There are ______ butterflies left.

K.OA.2

4) There were **5** bumble bees.
2 bumble bees flew away.
How many bumble bees now?

______ − ______

There are ______ bumble bees left.

K.OA.2

Do you understand? **?** ✓

Name: ______________________ Date: __________

Unit 8 Practice 4: I can find 1 more or 1 less.

1) **Sample:** Redraw the circles to show **1 more**. Write the new amount.

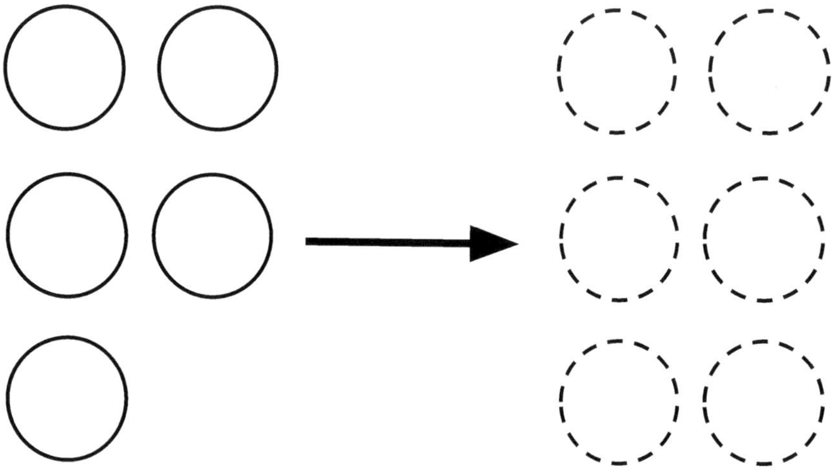

6

K.CC.4

2) Redraw the circles to show **1 less**. Write the new amount.

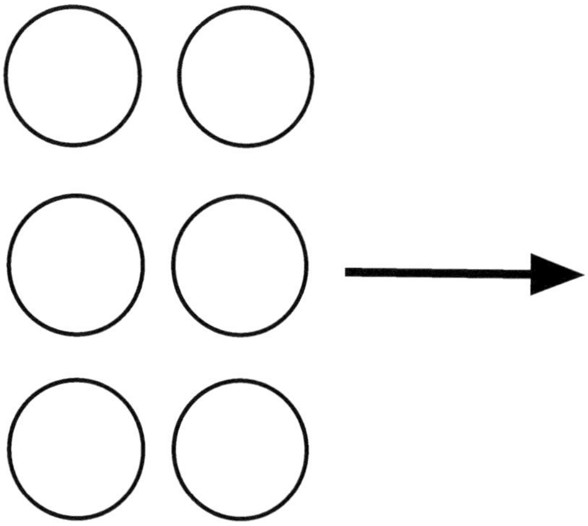

K.CC.4

Do you understand? **?** ✓

Affirmation: I have a powerful mind!

Name: ______________________ Date: ____________

Unit 8 Practice 4: I can find 1 more or 1 less.

3) Redraw the circles to show **1 more**. Write the new amount.

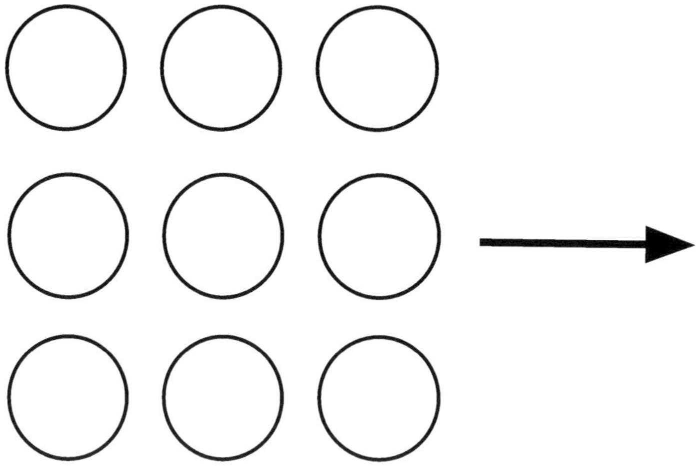

K.CC.4

4) Redraw the circles to show **1 less**. Write the new amount.

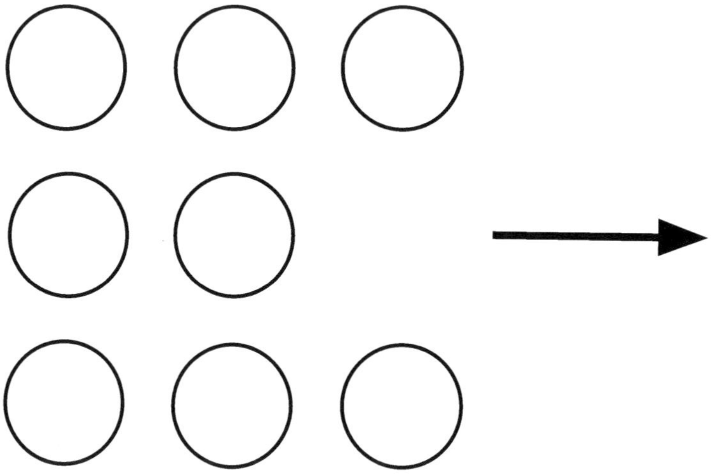

K.CC.4

Do you understand? **?** ✓

Name: ______________________ Date: ______________

Unit 8 Practice 5: I can write the number that comes next.

1) **Sample:** I can write the number that comes **next**.

5, 6, 7, 8, 9

K.G.4

2) I can write the number that comes **next**.

2, 3, 4, 5, ___

K.G.4

Do you understand? ? ✓

Name: ____________________ Date: __________

Unit 8 Practice 5: I can write the number that comes next.

3) I can write the number that comes **next**.

6, 7, 8, 9, ____

K.G.4

4) **Challenge:** I can write the number that comes **next**.

16, 17, 18, 19, ____

K.G.4

Do you understand? **?** ✓

Name: ______________________ Date: ____________

Unit 8 Practice 6: I can draw circles to match a number card.

1) **Sample:** Draw circles to match the number card.

3

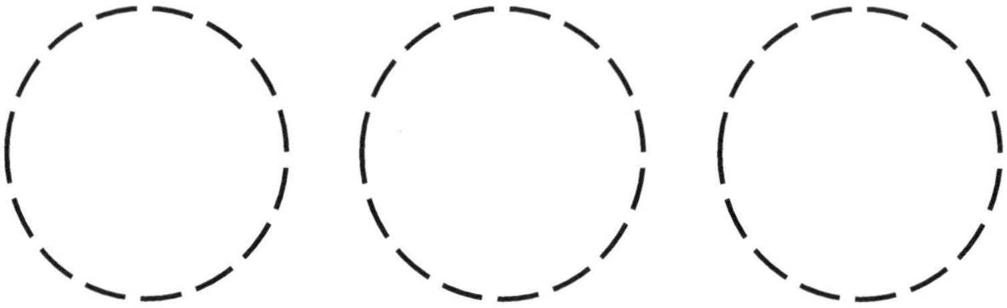

K.OA.1

2) Draw circles to match the number card.

6

K.OA.1

Do you understand? **?** ✓

Name: ______________________________ Date: ______________

Unit 8 Practice 6: I can draw circles to match a number card.

3 Draw circles to match the number card.

9

K.OA.1

4) Draw circles to match the number card.

12

K.OA.1

Do you understand? ? ✓

Name: ______________________ Date: ______________

Unit 8 Practice 7: I can write numerals and numeral names.

1) **Sample:** Write the numerals and numeral names to match the objects.

5 five

K.OA.1

2) Write the numerals and numeral names to match the objects.

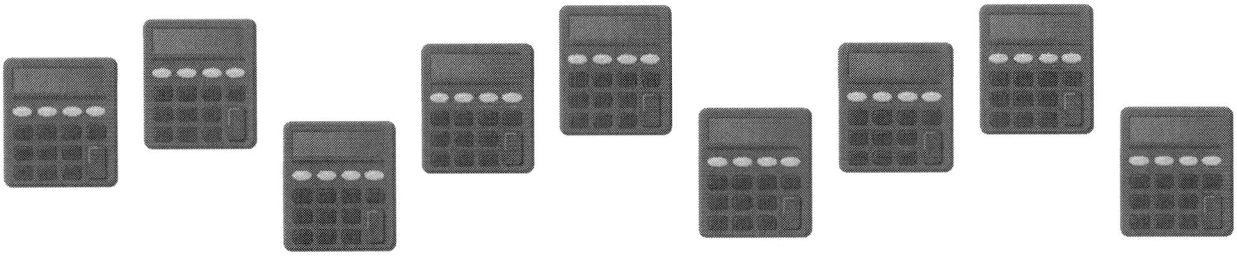

K.OA.1

Do you understand? **?** ✓

Affirmation: I have a powerful mind!

Name: ______________________ Date: ____________

Unit 8 Practice 7: I can write numerals and numeral names.

3 Write the numerals and numeral names to match the objects.

K.OA.1

4) **Challenge**: Write the numerals and numeral names to match the objects.

K.OA.1

Do you understand? **?** ✓

Affirmation: I am a mathematical thinker.

Name: ______________________________ Date: ______________

Unit 8 Practice 8: I can color counters to show an expression.

1) **Sample:** Color the counters to show the expression.

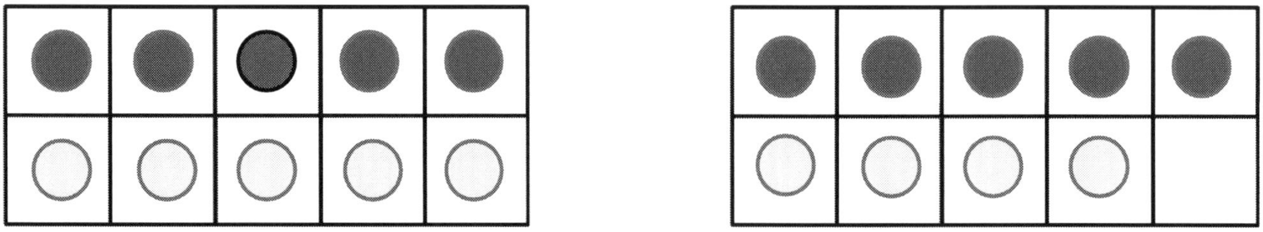

$$10 + 9$$

K.OA.1

2) Color the counters to show the expression.

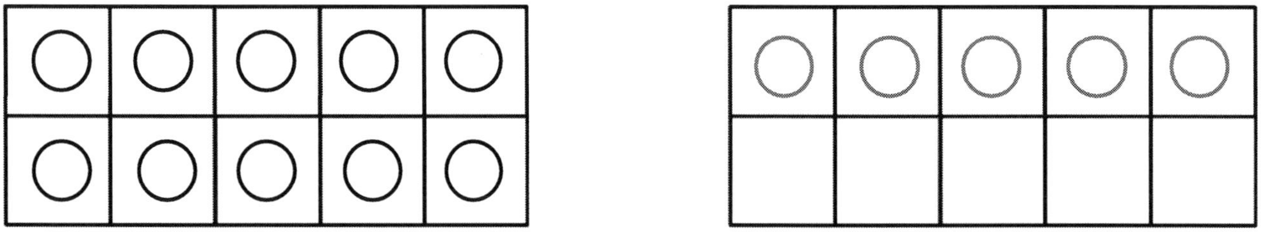

$$10 + 5$$

K.OA.1

Do you understand? **?** ✓

Name: ______________________________ Date: ______________

Unit 8 Practice 8: I can color counters to show an expression.

3 Color the counters to show the expression.

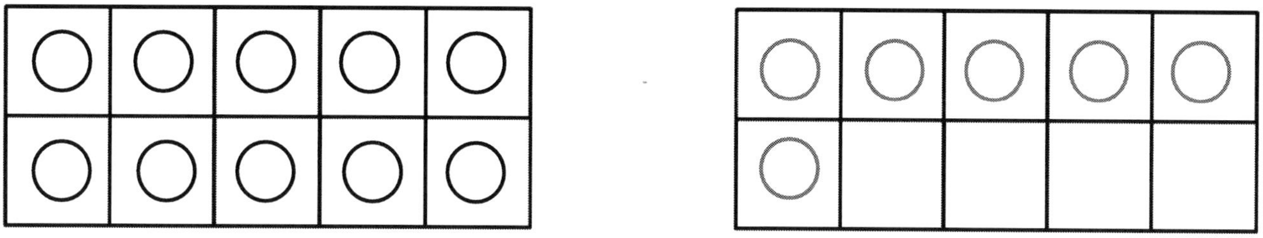

$$10 + 6$$

K.OA.1

4) Color the counters to show the expression.

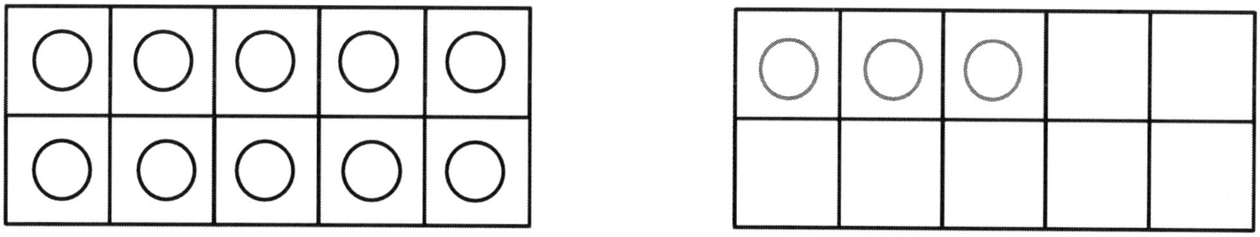

$$10 + 3$$

K.OA.1

Do you understand? **?** ✓

Name: ____________________ Date: ____________

Unit 8 Practice 9: I can answer questions.

1) **Sample:** How many letters can you find?

There are 6 letters.

K.CC.6

2) How many books can you find?

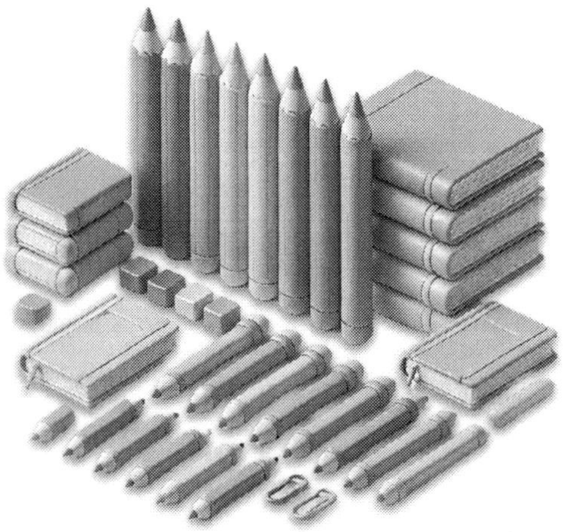

There are ________ books.

K.CC.6

Do you understand? ? ✓

Name: ______________________ Date: ____________

Unit 8 Practice 9: I can answer questions.

3) How many toy cars can you find?

There are ________ toy cars.

K.CC.6

4) How many pencils can you find that are standing up?

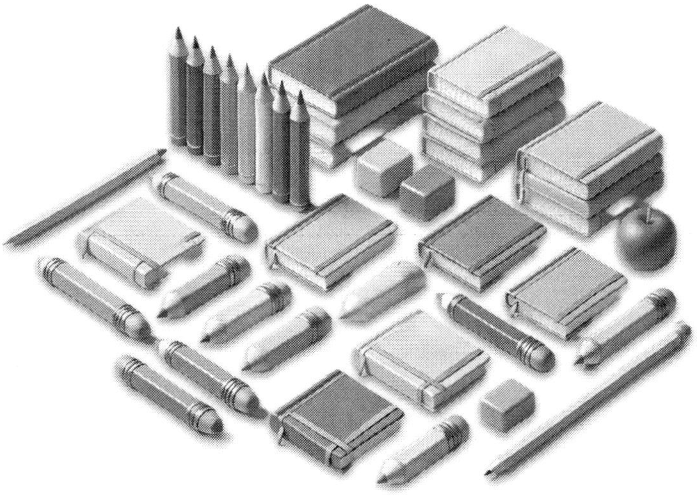

There are ________ pencils standing up.

K.CC.6

Do you understand? ? ✓

Affirmation: I am a mathematical thinker.

Name: ______________________________ Date: ______________

Unit 8 Practice 10: I can find the value of an expression.

1) **Sample:** Draw circles to show expression. Find the value of the expression.

3 + 2

5 circles

K.OA.1

2) Draw circles to show expression. Find the value of the expression.

3 + 4

____ circles

K.OA.1

Do you understand? ? ✓

Name: ______________________ Date: ____________

Unit 8 Practice 10: I can find the value of an expression.

3) Draw circles to show expression. Find the value of the expression.

$$5 - 2$$

_____

_ _ _ _

_____ circles

K.OA.1

4) Draw circles to show expression. Find the value of the expression.

$$7 - 4$$

_____

_ _ _ _

_____ circles

K.OA.1

Do you understand? **?** ✓

Name: ____________________ Date: ____________

Unit 8 Practice 11: I can write an equation to match a 10-frame.

1) **Sample:** Write an equation to match the 10-frame.

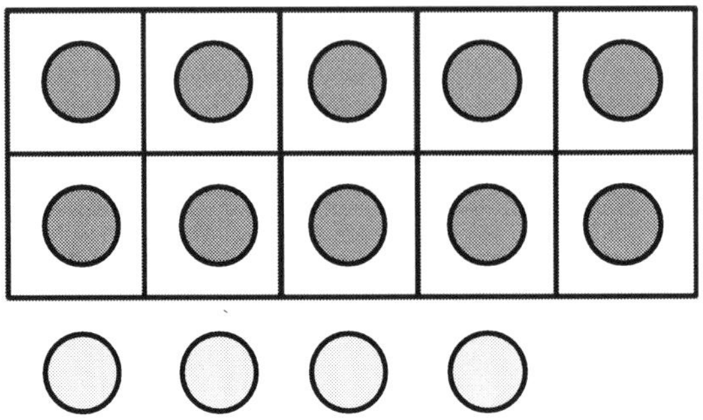

10 + 4 = 14

K.OA.1

2) Write an equation to match the 10-frame.

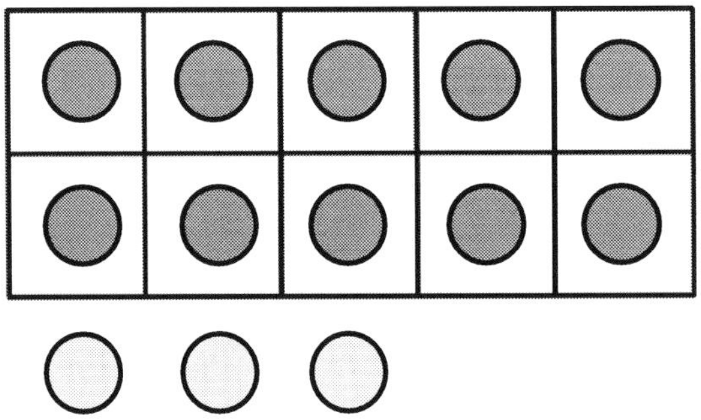

____ + ____ = ____

K.OA.1

Do you understand? ? ✓

Name: ______________________ Date: ____________

Unit 8 Practice 11: I can write an equation to match 10-frame.

3 Write an equation to match the 10-frame.

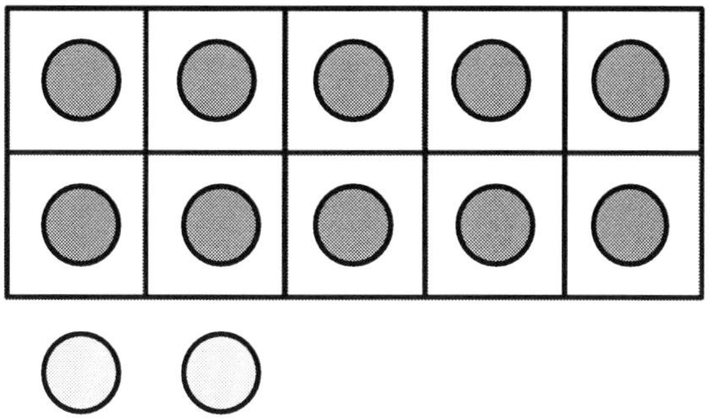

_____ + _____ = _____

K.OA.1

4) Write an equation to match the 10-frame.

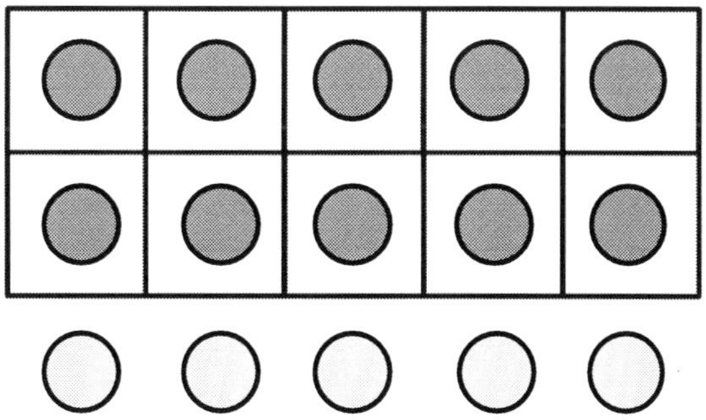

_____ + _____ = _____

K.OA.1

Do you understand? ? ✓

Name: ______________________ Date: ____________

Unit 8 Practice 12: I can tell how many using equations.

1) **Sample:** How many dots? Write how many.

4 + 6 = 10

10 + 2 = 12

12 dots

K.OA.1

2) How many dots? Write how many.

______ dots

K.OA.1

Do you understand? ? ✓

Affirmation: I am a mathematical thinker.

Name: ______________________ Date: ____________

Unit 8 Practice 12: I can tell how many using equations.

3 How many dots? Write how many.

_______ dots

K.OA.1

4) How many dots? Write how many.

_______ dots

K.OA.1

Do you understand? **?** ✓

Name: ____________________ Date: __________

Unit 8 Practice 13: I can write different expressions.

1) **Sample:** Write two expressions to match the domino.

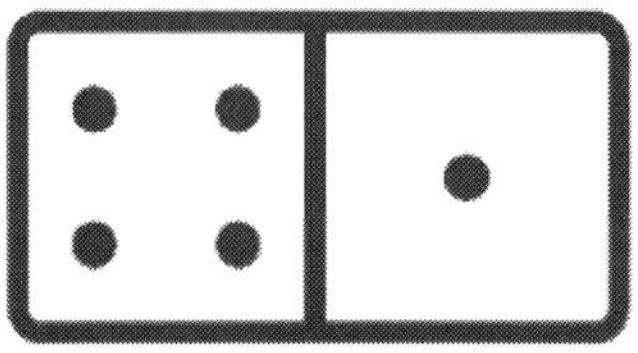

4 + 1 or 1 + 4

K.OA.1

2) Write two expressions to match the domino.

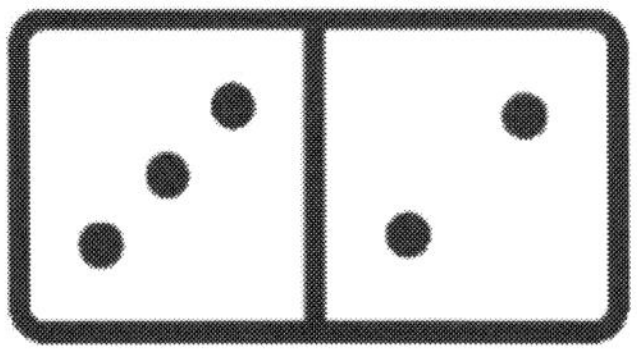

K.OA.1

Do you understand? ? ✓

Name: ____________________ Date: __________

Unit 8 Practice 13: I can write an expression.

3 **Challenge**: Write two expressions to match the domino.

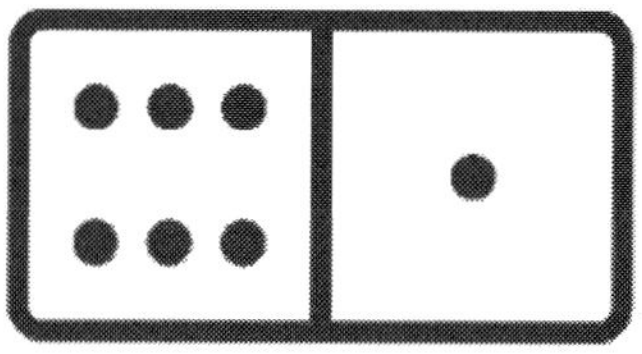

K.OA.1

4) **Challenge**: Write two expressions to match the domino.

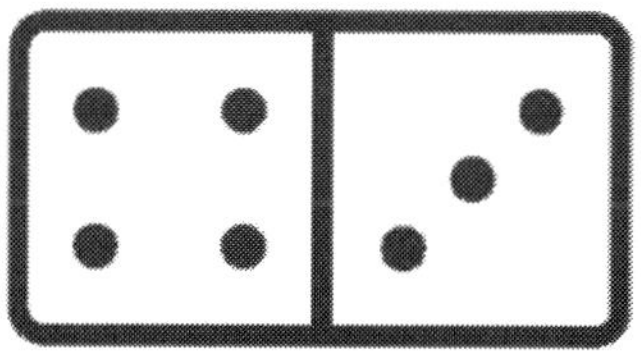

K.OA.1

Do you understand? ? ✓

Affirmation: We are all smart in different ways.

Name: ______________________ Date: ____________

Unit 8 Practice 14: I can break apart numbers.

1) **Sample:** Find two ways to break apart **5**. Draw circles to show your work.

5 = 2 + 3	5 = 4 + 1
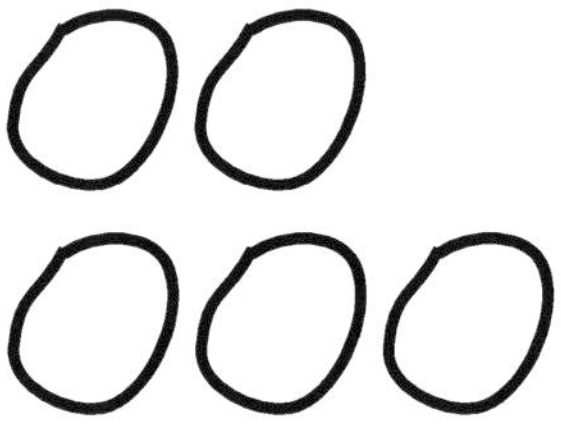	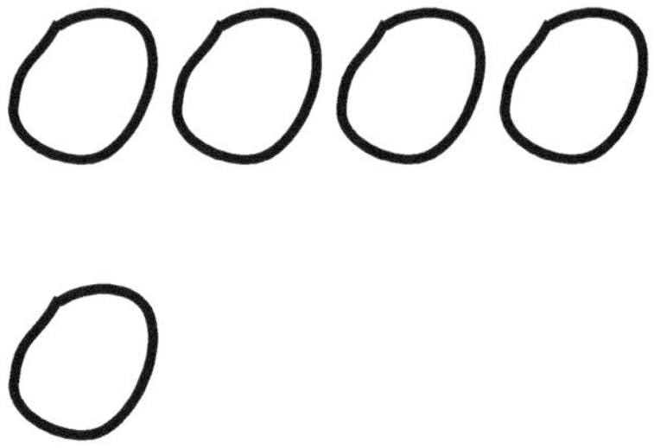

K.OA.3

2) Find two ways to break apart **4**. Draw circles to show your work.

4 = ____ + ____	4 = ____ + ____

K.OA.3

Do you understand? **?** ✓

Name: ______________________ Date: ____________

Unit 8 Practice 14: I can break apart numbers.

3) Find two ways to break apart **3**. Draw circles to show your work.

3 = _____ + _____	3 = _____ + _____

K.OA.3

4) **Challenge**: Find two ways to break apart **10**. Draw circles to show your work.

10 = _____ + _____	10 = _____ + _____

K.OA.3

Do you understand? **?** ✓

Affirmation: We are all smart in different ways.

Name: ______________________________ Date: ________________

Unit 8 Practice 15: I can tell which is more.

1) **Sample:** Check the expression that is more.

3 + 2 or 1 + 3

K.OA.2

2) Check the expression that is more.

2 + 2 or 3 + 2

K.OA.2

Do you understand? ? ✓

Name: ______________________ Date: ____________

Unit 8 Practice 15: I can tell which is more.

3) Check the expression that is more.

$1 + 4$ or $1 + 2$

K.OA.2

4) Check the expression that is more.

$0 + 5$ or $3 + 1$

K.OA.2

Do you understand? **?** ✓

Name: ______________________________ Date: ______________

Unit 8 Practice 16: I can make 5.

1) **Sample:** Complete the equation to make 5.
   Use drawing to show your thinking.

$$4 + 1 = 5$$

K.OA.2

2) Complete the equation to make 5.
   Use drawing to show your thinking.

$$3 + \_\_\_ = 5$$

K.OA.2

Do you understand? **?** ✓

Name: ____________________ Date: ____________

Unit 8 Practice 16: I can make 5.

3) Complete the equation to make 5.
Use drawings to show your thinking.

$6 - \_\_\_ = 5$

K.OA.2

4) Complete the equation to make 5.
Use drawings to show your thinking.

$7 - \_\_\_ = 5$

K.OA.2

Do you understand? ? ✓

Affirmation: We are all smart in different ways.

Name: ______________________________ Date: ______________

Unit 8 Practice 17: I can make 10.

1) **Sample:** Use the 10-frame to complete the equation.

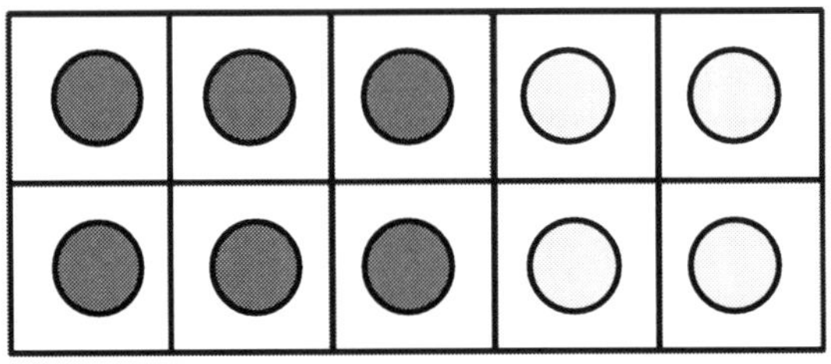

6 + 4 = 10

K.OA.1

2) Use the 10-frame to complete the equation.

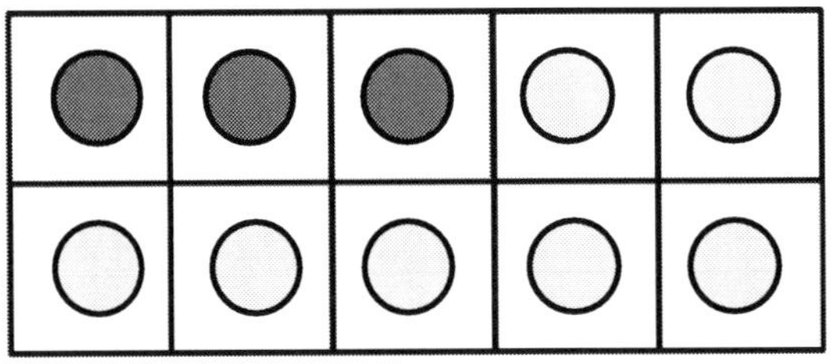

____ + ____ = 10

K.OA.1

Do you understand? ? ✓

Name: ______________________ Date: __________

Unit 8 Practice 17: I can make 10.

3) Use the 10-frame to complete the equation.

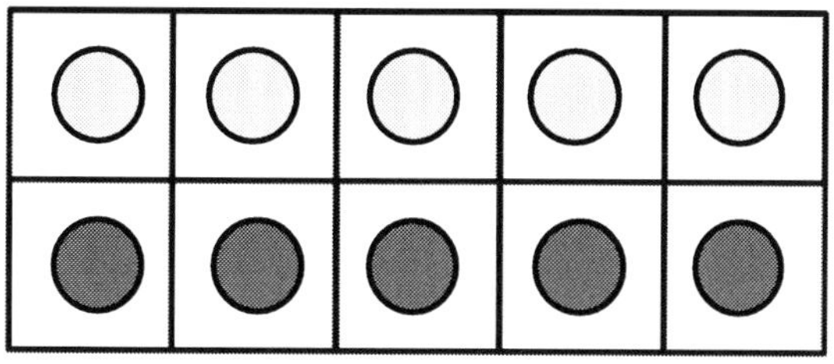

___ + ___ = ___

K.OA.1

4) **Make Your Own**: Shade in some circles. Use the 10-frame to complete the equation.

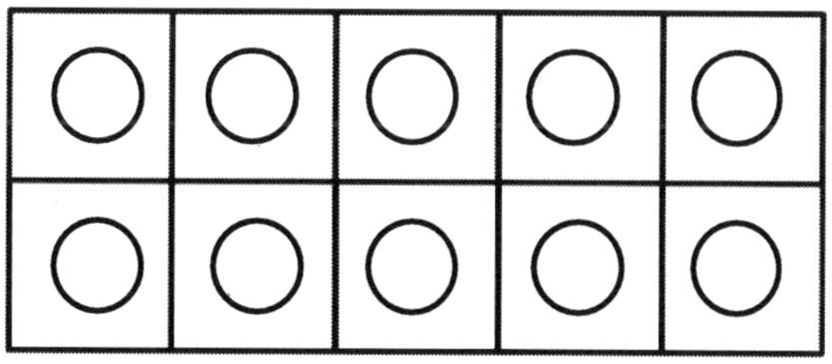

___ + ___ = 10

K.OA.1

Do you understand? ? ✓

Affirmation: Making mistakes is how I learn new things.

Name: ______________________________ Date: ______________

Unit 8 Practice 18: I can solve a story problem.

1) **Sample:** Raven saw **2** caterpillars.
Then he saw **8** more caterpillars.
How many caterpillars does Raven see?
Draw a picture to solve the story problem.

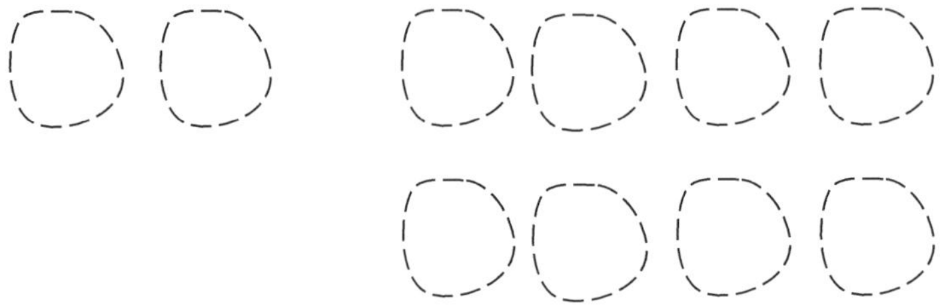

Raven saw <u>10</u> caterpillars.

K.OA.2

2) Pala saw **4** butterflies on a leaf.
Then she saw **6** butterflies on a stick.
How many butterflies does Pala see now?
Draw a picture to solve the story problem.

Pala saw ________ butterflies.

K.OA.2

Do you understand? ? ✓

Name: ______________________________ Date: ______________

Unit 8 Practice 18: I can solve a story problem.

3) Chen has **7** toy trucks.
Tala gave Chen **3** extra toy trucks.
How many toy trucks does Chen have now?
Draw a picture to solve the story problem.

Chen has __________ toy trucks now.

K.OA.2

4) Bilal has **5** teddy bears.
Salma gave Bilal **5** teddy bears.
How many teddy bears does Bilal have now?
Draw a picture to solve the story problem.

Bilal has __________ teddy bears now.

K.OA.2

Do you understand? **?** ✓

Name: ______________________ Date: ____________

Unit 8 Practice 19: I can make 10.

1) **Sample:** Fill in the equation to make 10.

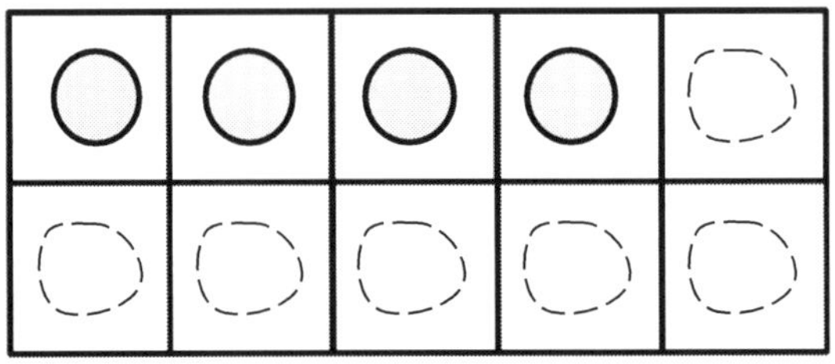

$$10 = 4 + 6$$

K.OA.4

2) Fill in the equation to make 10.

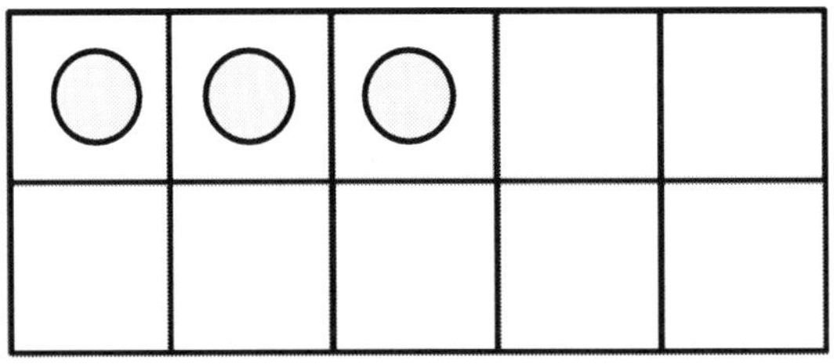

$$10 = 3 + \_\_\_$$

K.OA.4

Do you understand? **?** ✓

Name: ______________________ Date: ____________

Unit 8 Practice 19: I can make 10.

3) Fill in the equation to make 10.

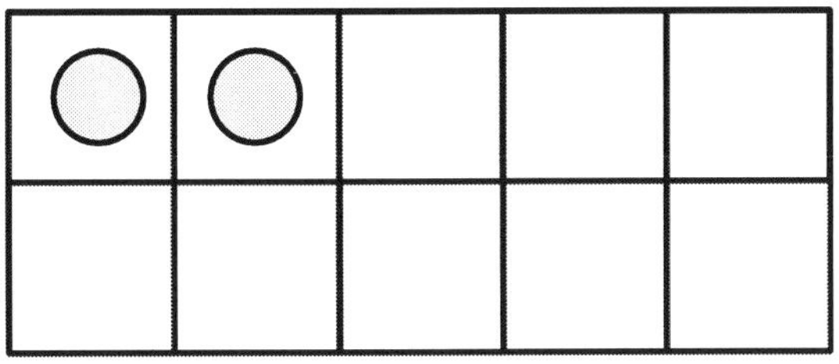

10 = 2 + ____

K.OA.4

4) Fill in the equation to make 10.

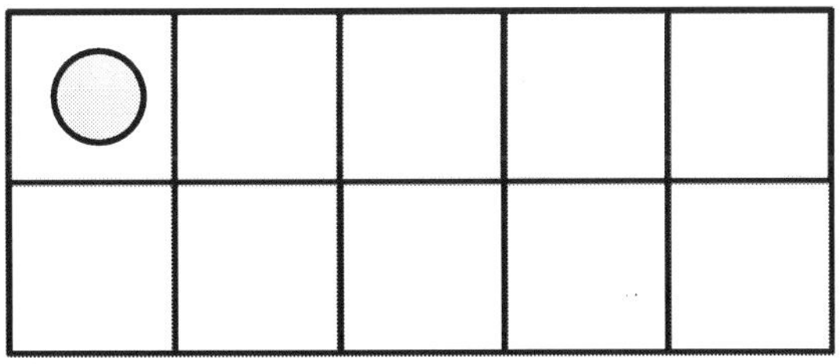

10 = 1 + ____

K.OA.4

Do you understand? ? ✓

Affirmation: Making mistakes is how I learn new things.

Name: ______________________ Date: ______________

Unit 8 Practice 20: I can tell more or fewer.

1) **Sample:** Circle **more than** or **fewer than** to make the sentence true.

There are **more than / fewer than** 10 ducks.

K.CC.5

2) Circle **more than** or **fewer than** to make the sentence true.

There are **more than / fewer than** 10 pigs.

K.CC.5

Do you understand? **?** ✓

Name: ______________________ Date: ____________

Unit 8 Practice 20: I can tell more or fewer.

3) Circle **more than** or **fewer than** to make the sentence true.

There are **more than / fewer than** 10 fish.

K.CC.5

4) Circle **more than** or **fewer than** to make the sentence true.

There are **more than / fewer than** 10 puppies.

K.CC.5

Do you understand? **?** ✓

Affirmation: Making mistakes is how I learn new things.

Name: ______________________________ Date: ______________

Unit 8 Practice 21: I can break apart numbers.

1) Color the cubes to show and complete the equation.

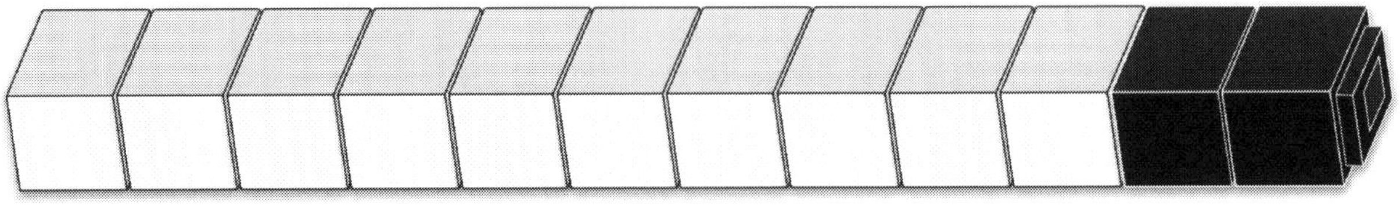

$12 = 10 + 2$

K.NBT.1

2) Color the cubes to show and complete the equation.

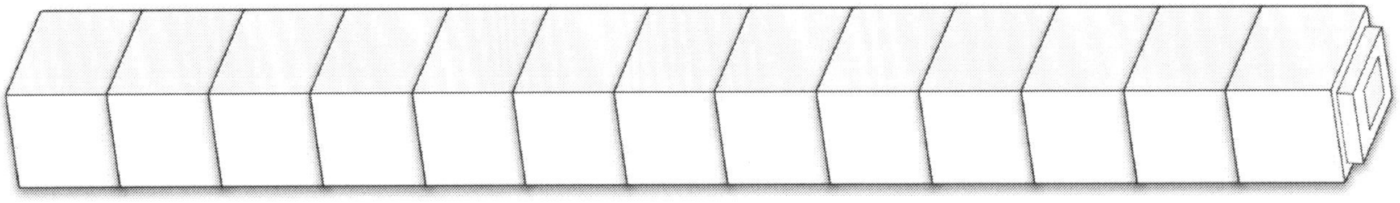

$13 = 10 + \_\_\_$

K.NBT.1

Do you understand? **?** ✓

Affirmation: Making mistakes is how I learn new things.

Name: ______________________________ Date: ______________

Unit 8 Practice 21: I can break apart numbers.

3) Color the cubes to show and complete the equation.

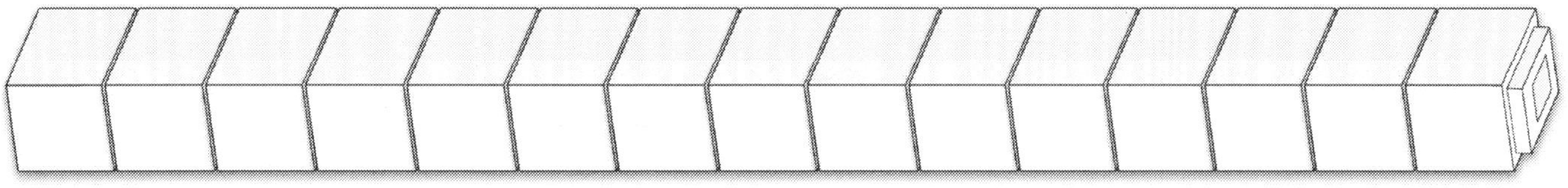

15 = 10 + ____

K.NBT.1

4) **Make Your Own**: Color the cubes to show and complete the equation.

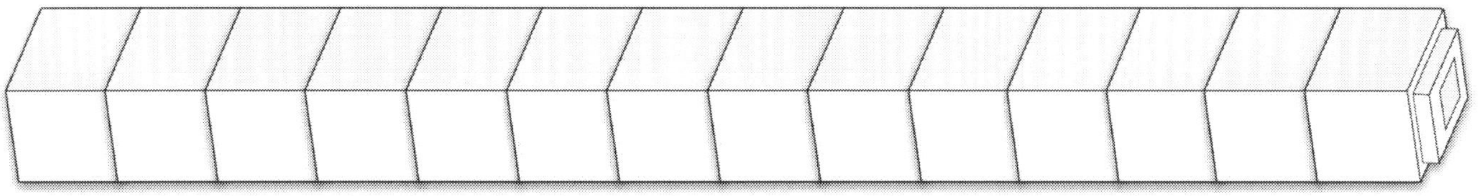

14 = ____ + ____

K.NBT.1

Do you understand? **?** ✓

Affirmation: I get better the more I practice.

Name: ______________________________ Date: ______________

I can practice grade level fluencies.

Set 1: Complete the equations.

1) 5 = 0 + _______

2) 5 = 1 + _______

3) 5 = 2 + _______

4) 5 = 3 + _______

5) 5 = 4 + _______

K.OA.1

Set 2: Complete the equations.

1) 5 + _______ = 10

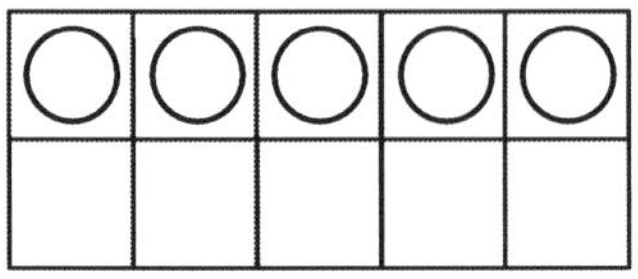

2) 6 + _______ = 10

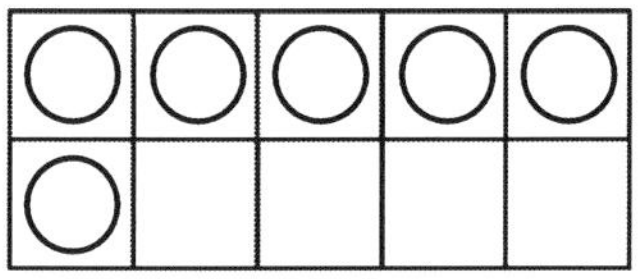

3) 7 + _______ = 10

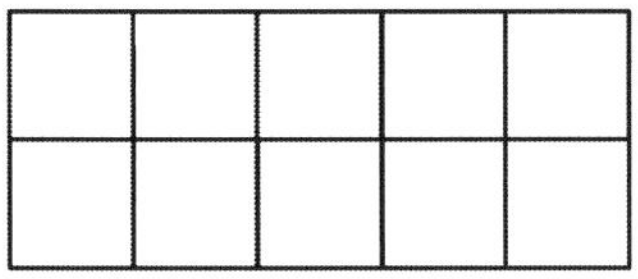

4) 8 + _______ = 10

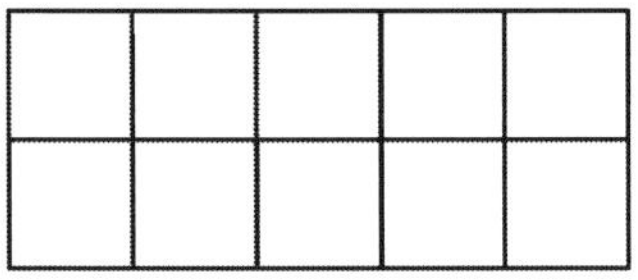

5) 9 + _______ = 10

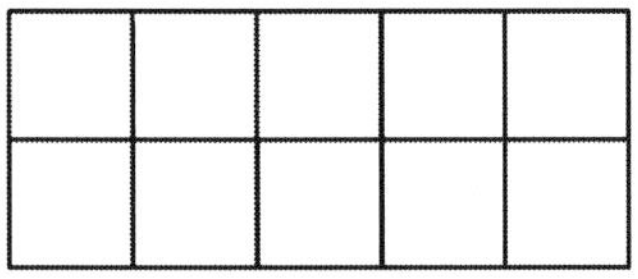

K.OA.1

Do you understand? ? ✓

Unit Reflection

Putting It All Together

Use this space to reflect on your understanding of the unit skills and concepts.

Skill/Concept	I can...	I need to work on...

Made in United States
Orlando, FL
05 November 2026

90230142R00092